SpringerBriefs in Agriculture

For further volumes:
http://www.springer.com/series/10183

Iván Francisco García-Tejero
Víctor Hugo Durán-Zuazo
José Luis Muriel-Fernández
Carmen Rocío Rodríguez-Pleguezuelo

Water and Sustainable Agriculture

Springer

Dr. Iván Francisco García-Tejero
Producción Ecológica y Recursos Naturale
IFAPA Centro "Las Torres-Tomejil"
Carretera Sevilla-Cazalla Km. 12,2
41200 Alcalá del Río Sevilla
Spain
e-mail: ivanf.garcia@juntadeandalucia.es

Dr. Víctor Hugo Durán-Zuazo
Producción Ecológica y Recursos Naturale
IFAPA Centro "Las Torres-Tomejil"
Carretera Sevilla-Cazalla Km. 12,2
41200 Alcalá del Río Sevilla
Spain
e-mail: victorh.duran@juntadeandalucia.es

Dr. José Luis Muriel-Fernández
Producción Ecológica y Recursos Naturale
IFAPA Centro "Las Torres-Tomejil"
Carretera Sevilla-Cazalla Km. 12,2
41200 Alcalá del Río Sevilla
Spain
e-mail: josel.muriel@juntadeandalucia.es

Dr. Carmen Rocío Rodríguez-Pleguezuelo
Suelos y Riegos
IFAPA Centro "Camino de Purchil"
Apartado de correos 2027
18080 Granada
Spain
e-mail: crocio.rodriguez@juntadeandalucia.es

ISSN 2211-808X
ISBN 978-94-007-2090-9
DOI 10.1007/978-94-007-2091-6
Springer Dordrecht Heidelberg London New York

e-ISSN 2211-8098
e-ISBN 978-94-007-2091-6

Cover design: eStudio Calamar, Berlin/Figueres

Printed on acid-free paper

Springer is part of Springer Science+Business Media (www.springer.com)

Contents

Water and Sustainable Agriculture

Abstract Irrigated agriculture is a vital component of total agriculture that supplies many of the fruits, vegetables, and cereal foods, the grains fed to animals used as human food, and the feed to sustain work animals in many parts of the world. Consequently, agriculture is the largest user of fresh water globally, and irrigation practices sometimes are biologically, economically, and socially unsustainable: wasting water, energy, and money; drying up rivers and lakes; reducing crop yields; harming fish and wildlife; and causing water pollution. There is an urgent need to reduce the amount of water used by producing more food, profits, livelihoods, and ecological benefits at lower social and environmental costs. To ensure sustainable irrigation the conditions under which crops grow should remain stable over a prolonged period. At the same time, soil degradation (salt accumulation), mining of the ground water aquifers, and the negative impact of drainage water on the downstream environment should be minimized. Water management should therefore balance the need of water for agriculture and the need for a sustainable environment. Consequently, irrigation has to be closely linked with water-use efficiency with the aim of boosting productivity and improving food quality, especially in those areas where problems of water shortages or collection and delivery are widespread. Certain improvements are possible through deficit-irrigation strategies, which have been proved to be sustainable, avoiding irrational application of water. Currently, agriculture is undergoing significant changes in innovative irrigation, fertilizer technology, and agronomic expertise. These elements constitute a vital platform for sustainable agricultural success and for preventing environmental impairment. The following book presents several processes and their link with environmental irrigation, balancing environmental protection with improved agricultural production. Thus, sustainable irrigation must be based on applying uniform and precise amounts of water, based on rational agricultural knowledge of the plant's water needs.

Keywords Water use · Climate change · Land management · Sustainable agriculture · Deficit irrigation

I. F. García-Tejero et al., *Water and Sustainable Agriculture*,
SpringerBriefs in Agriculture, DOI: 10.1007/978-94-007-2091-6_1,
© Iván Francisco García-Tejero 2011

1 Introduction

Water is indisputably one of the most precious of all natural resources and the limiting factor in economic and social development (Chenoweth 2008). Fresh-water resources globally are being over-exploited, polluted, and degraded and many systems are on the brink of collapse. At the same time, pressure has never been greater to provide drinking water as well as water for the economic development of a burgeoning world population. Water-resource management thus poses one of the great challenges for achieving ecologically sustainable development (Abu-Zeid 1998; Mariolakos 2007).

A growing world population is exerting increasing pressure on freshwater supplies. Global world population has exploded from 2.5 billion in 1,950–6 billion in 2000. By 2050, the world population could reach 10 billion or higher (UN 2007). In addition, to population pressures, there has also been an increase in water use per capita. The combination of these forces has led to worries about the adequacy of water supplies in the future.

Irrigation has allowed the world to overcome the potential food-supply problems associated with population growth. In many developing countries more than 90% of the water withdrawals are for irrigation (AQUASTAT 2005). In arid regions, irrigation is the essential for crop production. Similar for semi-arid and wet areas, irrigation boosts yields, attenuates the impact of droughts or, in the case of rice, minimizes weed growth. As stated by Bruinsma (2003), the average yields are generally higher under irrigated conditions as compared to rainfed agriculture. Globally only 18% of the cultivated area is irrigated (FAO 2005a), and 40% of food production comes from irrigated agriculture (UNCSD 1997). Both the water scarcity caused by using large amounts of water in irrigated agriculture and the importance of irrigation for crop production and food security induced several studies to quantify the different elements of the global water balance in space and time (Oki et al. 2001; Alcamo et al. 2003; FAO 2005b), to explore the importance of irrigated food production (Wood et al. 2000; Faures et al. 2002), and to assess the impact of climate change and climate variability in relation to global irrigation-water requirements (Döll 2002).

Worldwide, 278.8 million ha are equipped for irrigation, some 68% being is located in Asia, 17% in America, 9% in Europe, 5% in Africa, and 1% in Oceania. The largest areas of high irrigation density are found in northern India and Pakistan along the rivers Ganges and Indus, in the Hai He, Huang He, and Yangtze basins in China, and along the Nile river in Egypt and Sudan, among others. Irrigated lands produce 40% of the world's food supply, and the value of the output per cultivated area is extremely high (Dregne and Chou 1992). In addition, there is some evidence that the high productivity of irrigated agriculture has slowed the rate of deforestation. Agriculture is one of the primary reasons for deforestation in many countries, and increased yields (a change at the intensive margin) have decreased the need to expand the total land under cultivation (a change at the extensive margin). This is particularly important in mountainous regions because the steep slope of this land not only makes it difficult for crop production; it also increases its value in conservation against forest fires. Many changes in ecosystems increase their vulnerability, and therefore it is essential to adopt soil-conservation measures and to decrease erosion and runoff into water bodies (Schröter et al. 2005).

Worldwide water consumption in 2000 was 4–5 times that of 1,950 levels. Most of the obvious sources of water have been developed, and many that remain are marginal at best. Recently, many countries have increased their understanding of the importance of freshwater for environmental services, such as ecosystem health, as well as the environmental costs of water projects, such as habitat destruction (Hubbard et al. 2005; Ojeda et al. 2008; De Groot and Hermans 2009).

Soil salinity is a problem on irrigated arid lands, both reducing productivity and forcing land out of production. In many places with insufficient surface-water supplies, groundwater is used as a substitute. While the availability of groundwater has benefited the global food supply, its use as an input has progressed in an unsustainable manner. As much as 8% of food crops grown on farms use groundwater faster than the rate at which aquifers are replenished. Using irrigation in a more efficient manner will be necessary to protect water sources while still meeting objectives of food security (Huang et al. 2006; Turral et al. 2010; De Fraiture et al. 2010).

Globally, the world has enough water; however, the water is unevenly distributed. Industries and households are increasingly demanding water at the expense of agriculture. Agriculture uses approximately 70% of the total amount of water withdrawn to supply our current food needs. By 2030 the Food and Agriculture Organisation estimates that we will require 60% more food to feed the world's ballooning population. If agricultural production is to be sustainable, water resources must be used more efficiently while still increasing agricultural productivity. By using a range of agricultural techniques and technologies such as modern irrigation techniques, integrated pest management and biotechnology, farmers can produce higher yields with higher quality produce while making the most of precious water supplies (Burke et al. 1999; Playán and Mateos 2006; Metzidakis et al. 2008).

The plant-science industry is ready to address the challenge of increasing agricultural productivity, providing a range of products from seeds, chemical crop-protection products and biotechnology products. The industry strives to improve the quality and yields of their produce using the same amount of land by gaining access to knowledge and information that can help them make the correct choices to improve their livelihoods in a manner compatible with sustainable agriculture (Mannion 1995a; Lyson 2002; Huang et al. 2004).

Another well-known, often quoted fact is that irrigation uses 70% of the water in human use, with an efficiency of 40–50%. That is an increase of that efficiency to 80–90% with drip irrigation that would save half the water and would be the obvious solution to the world's water problems (Heermann ét al. 1990).

More accurately expressed, irrigated agriculture is responsible for some 70% of all blue water withdrawn for human use. At the turn of the century, some 10%

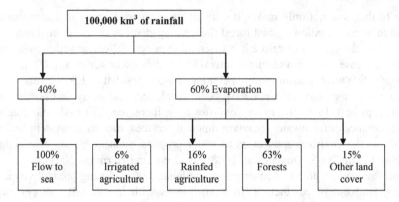

Fig. 1 Assessment of evapotranspiration from different land covers (source International Water Management Institute)

of all blue water, about 4,000 km^3, was withdrawn for human use (Rijsberman and De Silva 2006).

In other words, a bit less than 3% of all water (green and blue) was withdrawn for irrigated agriculture. According to International Water Management Institute (www.iwmi.cgiar.org) irrigated agriculture was responsible for some 6% of all evapotranspiration, while a further 16% was used consumptively by rainfed agriculture (Fig. 1). By contrast, natural ecosystems, from forests to wetlands, still use far more water than agriculture does.

It is also a misconception that irrigation efficiency can be increased, i.e. from 40 to 80%, which then would lead to major savings. In this sense, Seckler et al. (2003) pointed out that "irrigation efficiency" is a confusing term that has been defined in too many ways. Whether increased irrigation efficiency in a farmer's field does indeed save water depends heavily on the fate of the return flow (the drainage water and recharge to the groundwater). At the scale of the farmer's field, water productivity can be measured in units of output (the crop per drop) or as the value in monetary units. At the watershed scale, water productivity should be understood in the widest possible sense, *i.e.* including crop, livestock and fishery yields, wider ecosystem services and social impacts such as health, together with the systems of resource governance that ensure equitable distribution of these benefits (Ali and Talukder 2008; Haileslassie et al. 2009). In this sense, Zoebl (2006) pointed out that a split-up of domains is needed to properly define water-use efficiency for different perspectives: a universal or single menu cannot be applicable due to the technical and socio-economic diversity of issues and challenges.

In Europe the directive 2000/60/EC of the European Parliament and the Council of 23 October 2000 establishing a framework for Community action in the field of water policy "Water Framework Directive" entered into force on 22 December 2000. The directive as one of the most substantial pieces of European Union water legislation combines the until then rather fragmented European Union water law (large number of directives dealing only with special aspects of water management such as waste water, dangerous substances, drinking water, etc.) in order to ensure sustainable water management.

Since the 1960s, the Common Agricultural Policy in the European Union has played a major role in supporting farmers. However, the present framework of this policy will cease in 2013. By that time, when the financial support and payments to producers are scheduled to decrease, the Water Framework Directive should have been completely implemented (by 2012). By requiring the full-cost principle, the implementation of the Water Framework Directive will probably lead to a water-price increase. The joint effects of water policies and decrease in Common Agricultural Policy payments will thus affect the viability of farming systems, especially in the case of irrigated farms. To date, scientists have partially investigated the impact of expected changes in parameters such as water price or Common Agricultural Policy payments on economic, social, and environmental sustainability of irrigated agricultural systems. In this context, Bartolini et al. (2007) for Italy reported that there is a clear trade-off between reducing the negative environmental impacts of agriculture and maintaining the livelihood of the sector. Overburdening farmers with increases in water prices could strongly influence the sustainability of the sector. Nevertheless, some specific crops such as non-intensive crop systems (rice and cereals) could sufficiently adapt to increased water prices (e.g. by reducing their water use or improving their irrigation system) Therefore, in these cases the water pricing could be a good economic mechanism in order to provide incentives for saving water. Furthermore, the results of this study highlight the need for more integrated analysis when setting such water policies. The economic viability of farming systems should also be taken into account in the design of policies and in particular of regulations related to water.

Water stress influences crop growth and productivity in many ways (Wanjura et al. 1990; Irmak et al. 2000). Most of the responses have a negative impact on yield but crops have different and often complex mechanisms to react to water shortages. Several crops and genotypes have developed different degrees of drought tolerance, drought resistance or compensatory growth to deal with periods of stress. The highest crop productivity is achieved for high-yielding varieties with optimal water supply and high soil-fertility levels, but under conditions of limited water supply, crops will adapt to water stress and can produce well with less water. In this sense of improving water productivity, there is growing interest in deficit irrigation, an irrigation practice whereby water supply is reduced below maximum levels and mild stress is allowed with minimal effects on yield (Ali et al. 2007; Ali and Talukder 2008). Under conditions of scarce water supply and drought, deficit irrigation can lead to greater economic gains than maximizing yields per unit of water for a given crop; farmers are more inclined to use water more efficiently, and more water-efficient cash-crop selection helps optimise returns (Geerts and Raes 2009; Blum 2009). However, this approach requires precise knowledge of crop response to water, as drought tolerance varies considerably by species, cultivar, and growth stage.

In the many regions, soil moisture is generally limited and crop growth is stressed by drought during the growing season, resulting in decreased and unsustainable crop yields. Particularly for the arid and semi-arid areas, rainfall is low, making crop growth dependent on irrigation. However, in sloping fields, most of the crops are grown under rainfed conditions. Recent research has shown some practical techniques of rainwater harvesting (Li et al. 2002; 2007; Tian et al. 2003) but the cost to prevent runoff and harvest rainwater on sloping lands can be very high (Zhang et al. 2004). On the other hand, it may not be practical to irrigate fields on a large scale using only harvested rainwater.

Sustainable development requires pragmatic management of land–water resources through positive and realistic science-based planning that balances the ecosystem's carrying capacity with respect to human expectations (Sophocleous 2000; Melloul and Collin 2003; Mariolakos 2007). The aim must be not only environmental harmony, but also long-term sustainability of natural resources with economic efficiency as its intent so as to meet the needs of the current generation without compromising the ability of future generations to meet their own needs (Costanza 1995).

Conceptual understanding and operational procedures applicable to improving system management and performance, and management changes in individuals and organizations are the changes necessary to meet the urgent needs in irrigated agriculture. Urgent needs relate to productivity, water scarcity, and managing the environment.

There is an urgent need to reduce the amount of water used for agriculture by producing more food with high quality, profit, livelihoods, and ecological benefits at less social and environmental costs per unit of water used. Water productivity defined in physical terms is the ratio of the mass of agricultural output to the amount of water used. In an economic sense, water productivity reflects the value derived per unit of water applied. Improving physical water productivity in irrigated and rainfed agriculture reduces the need for additional water and is thus a critical response to increasing water scarcity (Molden et al. 2007).

This review shows some urgent changes needed and defines the conceptual and operational strategies that can address the needs and accomplish the sustainability of irrigation systems. Research needs to provide methods for improved productivity, for more effective use of water supplies while making available additional irrigation water, and for approaching environmental sustainability in irrigated agriculture.

2 Water and Agriculture

Agriculture represents the first, traditional life-supporting economic sector closely linked to establish cultural and ethical values of land and water on which traditional societies are built. Water in agriculture is largely associated with irrigation. The worldwide area equipped with irrigation expanded from 139 million ha in 1961–277 million ha in 2003 (FAO 2007). According to Bhattarai et al. (2007), the investments in irrigation have increased rural incomes, resulting in greater demands for nonfarm goods and services. Over one-third of the world's food is now produced on the irrigated 17% of the world's croplands. There are currently some 280 million ha of irrigated area, almost three quarters of it in developing countries. Only four countries contain half the world's irrigated land: China, India, United States of America, and Pakistan. The remaining two-thirds of all food is being produced on a five times larger rainfed area amounting to 1,250 million ha.

Plants require water for photosynthesis, growth, and reproduction. Water used by plants is non-recoverable, because some water becomes a part of the plant chemically and remainder is released into the atmosphere. The processes of CO_2 fixation and temperature control require plants to transpire enormous amounts of water. Many crops transpire water at rates from 600 to 2,000 L kg^{-1} of dry matter (Klocke et al. 1996; USDA 1997; USDA-NASS 1998; Snyder 2000). According to Schlesinger (1997) the average global transfer of water into the atmosphere from the terrestrial ecosystems by vegetation transpiration is estimated to be about 64% of all precipitation that falls to earth.

In the other hand, irrigation requires a significant expenditure of fossil energy both for pumping and delivering water to crops. As stated by Hodges et al. (1994), annually in the U.S., about 15% of the total energy expended for all crop production is used to pump irrigation water. Overall the amount of energy consumed in irrigated crop production is substantially greater than for rainfed crops. For example, irrigated wheat requires more than three times of energy than rainfed wheat. Concretely, about 4.2 million kcal ha^{-1} yr^{-1} are the required energy input for rainfed wheat, while irrigated wheat requires 14.3 million kcal ha^{-1} yr^{-1} to apply an average of 5.5 ml of water (Pimentel et al. 2002a). Delivering the 10 ml of irrigation water needed by a hectare of irrigated corn from surface water sources requires the expenditure of about 880 kW ha^{-1} of fossil fuel (Batty and Keller 1980). By contrast, when irrigation water must be pumped from a depth of 100 m, the energy cost increases up to 28,500 kWh ha^{-1}, or more than 32 times the cost of surface water (Gleick 1993). Therefore, the costs of irrigation for energy and capital are significant. Farmers must not evaluate only the economic cost of developing irrigated land, but must also consider the annual costs of irrigation pumping. The large quantities of energy required to pump irrigation water are significant considerations both from the standpoint of energy and water resource management. According to Pimentel et al. (2002b), about 8 million kcal of fossil energy are expended for machinery, fuel, fertilizers, pesticides, and partial (15%) irrigation, to produce one hectare of rainfed corn. In contrast, if the corn crop were fully irrigated, the total energy inputs would rise to nearly 25 million kcal ha^{-1} (2,500 L of oil equivalents) (Gleick 1993).

The minimum soil moisture essential for crop growth varies. In this context, Broner (2002) reported that potatoes in the U.S. require 25–50%, alfalfa 30–50%, and corn 50–70%, while rice in China is reported to require at least 80% soil moisture according to Zhi (2000).

The water required by food and forage crops ranges from 600 to 3,000 L kg^{-1} d.m. of crop yield. A hectare of U.S. corn, with a yield of approximately 9,000 kg ha^{-1}, transpires about 6 million L ha^{-1} during the growing season (Benham et al. 1999;

Palmer 2001), while an additional 1–2.5 million L ha^{-1} of soil moisture evaporate into the atmosphere (Desborough et al. 1996). This means that about 800 mm (8 million L ha^{-1}) of rainfall are required during the growing season for corn production. A hectare of high-yielding rice requires approximately 11 million L ha^{-1} of water for an average yield of 7 t ha^{-1} (Snyder 2000). On average, soybeans require about 5.8 million L ha^{-1} of water for a yield of 3 t ha^{-1} Benham et al. 1999. Meanwhile, wheat, which produces less plant biomass than either corn or rice, requires only about 2.4 million L ha^{-1} of water for a yield of 2.7 t ha^{-1} (USDA 1997).

World agriculture consumes approximately 70% of freshwater withdrawn per year (UNESCO 2001). About 17% of the world's cropland is irrigated but produces 40% of the world's food (FAO 2002b). Worldwide, the amount of irrigated land is slowly expanding, even though salinization, water logging, and silting continue to diminish productivity (Gleick 2002). Despite a small annual increase in total irrigated areas, the per capita irrigated area has been declining since 1990, due to rapid population growth (Postel 1999; Gleick 2002). Specifically, global irrigation per capita has declined nearly 10% during the past decade (Postel 1999; Gleick 2002), while in the U.S. irrigated land per capita has remained constant at about 0.08 ha (USDA 2001).

The efficiency varies with irrigation technologies and most common irrigation methods, flood irrigation and sprinkler irrigation, frequently waste water. By contrast, the use of more focused application methods, such as "drip" or "micro-irrigation" have found favour because of their increased water use efficiency. Drip irrigation delivers water to individual plants by pipes and uses less water than surface irrigation. In addition, to conserving water, drip irrigation reduces the problems of salinization and waterlogging (Tuijl 1993). Although drip systems achieve up to 85% water use efficiency, they are expensive, may be energy intensive, and require clean water to prevent the clogging of the fine delivery tubes (Heermann et al. 1990). With rainfed crops, salinization is not a problem because the salts are naturally flushed away. However, when irrigation water is applied to crops and returns to the atmosphere via plant transpiration and evaporation, dissolved salts concentrate in the soil, where they inhibit plant growth. Bouwer (2002) pointed out that the practice of applying about 10 million L ha^{-1} yr^{-1} of water, results in approximately 5 t ha^{-1} of salts being added to the soil. The salt deposits can be flushed away with added fresh water but at an important cost. Worldwide, approximately half of all existing irrigated soils are adversely affected by salinization (Hinrichsen et al. 1998). According to Pimentel et al. (2004), each year the amount of world agricultural land affected by salinized soil is estimated to be 10 million ha. Also, drainage water from irrigated cropland contains large quantities of salt.

Waterlogging is another problem associated with irrigation (Bowonder et al. 1986; Smedema 1990). Over time, seepage from irrigation canals and irrigated fields cause water to accumulate in the upper soil levels. Due to water losses during pumping and transport, approximately 60% of the water intended for crop irrigation never reaches the crop (Wallace 2000). In the absence of adequate drainage, water tables rise in the upper soil levels, including the plant root zone, and crop growth is impaired. In India, waterlogging adversely affects 8.5 million ha of cropland and results in the loss of as much as 2 million t yr^{-1} of grain every year (ICAR 1999).

On the other hand water erosion adversely affects crop productivity by reducing the availability of water, diminishing soil nutrients, soil biota, and organic matter, and also decreasing soil depth (Pimentel and Kounang 1998; Durán et al. 2006, 2010). In this context, Guenette (2001) reported that the reduction in the amount of water available to the growing plants is considered the most harmful effect of erosion, because eroded soil absorbs 87% less water by infiltration than uneroded soils. Soybean and oat plantings intercept approximately 10% of the rainfall, whereas tree canopies intercept 15–35%.

Surface runoff, which carries sediments, nutrients, and pesticides from agricultural fields, into surface and groundwater, is the leading cause of non-point source pollution in many agricultural lands of the world (Liu et al. 2005; Duchemin and Hogue 2009). Thus, soil erosion is a self-degrading cycle on agricultural land. As erosion removes topsoil and organic matter, water runoff is intensified and crop yields decrease. The cycle is repeated again with even greater intensity during subsequent rains.

Increasing soil organic matter by applying manure materials can improve the water-infiltration rate by as much as 150% (Guenette 2001). In addition, using vegetative cover, such as intercropping and grass strips, helps slow both surface runoff and erosion (Lal 1993; Durán et al. 2008, 2009a). For example, when silage corn is interplanted with red clover, surface runoff can be decreased by 87% and soil loss by 78% (Wall et al. 1991). Reducing water runoff in these and other ways is an important step in increasing water availability to crops, conserving water resources, decreasing non-point source pollution, and ultimately decreasing water shortages.

According to Roose (1996), planting trees to serve as shelter belts between fields reduces evapotranspiration from the crop ecosystem by up to 20% during the growing season, thereby reducing non-point source pollution, and increases some crop yields, such as potatoes and peanuts (Snell 1997). If soil and water conservation measures are not implemented, the loss of water for crops via soil erosion can amount to as much as 5 million L ha^{-1} yr^{-1} (Pimentel and Kounang 1998).

In terms of agricultural yield, the production of animal protein requires significantly more water than the production of plant protein (Pimentel and Pimentel 2003). As stated by Solley et al. (1998), livestock directly uses only 2% of the total water applied by agriculture, the water input for livestock production being substantial because water is required for the forage and grain crops. In the U.S. each year, about of 253 million t of grain are fed to livestock, requiring a total of 250×10^{12} L of water (USDA-NASS 2002). Worldwide grain production specifically for livestock requires nearly three times the amount of grain that is fed U.S. livestock and three times the amount of water used in the U.S. to produce the grain feed (Seglken 1997, Earth Policy Institute 2002). Along this line, the production of 1 kg of chicken requires 3,500 L of water, while producing 1 kg of sheep requires approximately 51,000 L of water, counting the 21 kg of grain and 30 kg of forage needed to feed these animals (USDA 2001). U.S. agricultural production is projected to expand in order to meet the increased food needs of its population, which is projected to double in the next 70 years (USBC 2001). The food situation is expected to become more critical in developing countries (Rosengrant et al. 2002). Increasing crop yields necessitates a parallel increase in freshwater utilization in agriculture. Therefore, increased crop and livestock production during the next five to seven decades will significantly increase the demand on all water resources, especially in the western, southern, and central United States (USDA 2001), as well as in many regions of the world with low rainfall.

Water for agriculture covers a wide range of consumptive and non-consumptive water uses in all the agricultural sub-sectors related to ethical conflicts and significant social, economic, and environmental issues. Agricultural water represents the dominant water use in the form of pumping for irrigation or rainwater and soil moisture in croplands and forests. Evaporation from freshwater bodies and wetlands is important for biodiversity and inland and marine fisheries. Irrigation uses about 70% of total globally extracted water volumes, estimated at 6,800 km^3 yr^{-1}, while total agricultural use represents about 92% of total uses of flowing water and rainwater (25,000 km^3 yr^{-1}; Appelgren 2004).

Table 1 shows various categories of agricultural water and its use for crops (irrigated, rainfed, and dryland), livestock, fisheries, and forestry sub-sectors according to FAO (2000). Agricultural demands represent actual current use of rainfall, soil moisture, and flowing waters for agricultural production. Above all, agriculture provides the food for the world's populations under both rainfed and irrigated agricultural systems. In a wider perspective, agriculture is not only the main consumer of water but also a crucial factor shaping important terrestrial and freshwater resources that form part of necessary life-supporting eco-system services. Agriculture has also become a critical cause and a source of water pollution that has also upset the nutrition cycle in the watercourses and soil–water systems and rendered the water unsuitable or less valuable for other water uses.

Higher levels of nitrogen application lead to more nitrate (NO_3) leaching to groundwater, streams, and rivers. Agriculture has become the largest source of nitrogen and phosphorus in waterways and coastal zones (Carpenter et al. 1998; Bennett et al. 2001). Contamination of groundwater is common in agricultural regions around the world (Matson et al. 1997). High NO_3–N concentrations in drinking water constitute a human health hazard and challenge the health of natural systems. Eutrophication of estuaries and other coastal marine environments can cause low- or zero-oxygen conditions, leading to loss of fish and shellfish and to algae blooms that are toxic to fish (Nixon 1995; Howarth et al. 1996).

Nitrogen fertilizer is the largest input of nitrogen and accounts for a major portion of leaching losses from cereal crops (Webster et al. 1999). The synchronization of nitrogen supply and demand, without excess or deficiency, is the key to optimising tradeoffs between enhancing water productivity to minimize water use and limiting NO_3–N leaching to reduce adverse effects on water quality (Dinar et al. 1991; Moreno et al. 1996a; Lemaire et al. 2007). Setting the research

Table 1 Water uses in agriculture and its impact (FAO 2000)

Agricultural sub-sector	Rivers and groundwater source	Rainfall water and soil moisture use	In-stream use and impacts	Virtual water trade changes in stock, environmental degradation
Irrigation	12% of cultivated area	Supplementary Irrigation/water harvesting; non-rival uses	Non-point source river pollution	Food trade; food aid
Rainfed crops	Supplementary irrigation/ water harvesting; rural water supplies	88% of cultivated area; cash export crop production	Non-point source river pollution	Food trade; food aid; land degradation
Livestock	Livestock watering supplies; production based on irrigated fodder	Range; grazing land management	Grazing of seasonally flooded lowland	Reduction and expansion in stock; food trade; food aid; land degradation
Fisheries	Pond aquaculture	–	Capture fisheries; farming in natural water bodies	Food trade; food aid
Forestry	Forest irrigation	Annual consumptive use (500–1,500 mm) reduced runoff, improved low-inflows	Impact on sedimentation and floods	Reduction in stock, export/import of timber and fuel wood

agenda and developing effective management practices to meet this challenge require a quantitative understanding of the current levels of nitrogen-use efficiency and NO_3–N losses.

Smil (1999) estimates total the nitrogen input to the world's cropland at 169 million t N yr^{-1}. Inorganic fertilizer supplies 46% of the total, biological fixation from legumes and other nitrogen-fixing organisms 20%, atmospheric deposition 12%, animal manure 11%, and crop residues 7%. NO_3 and phosphorus enrichment is degrading water quality, promoting eutrophication processes. About the 80% of European surface waters exceed the European Commission's drinking water standard for 50 mg NO_3 L^{-1} (Molenat and Gascuel 2002). For example, in England and Wales, 28% of rivers exceed nitrogen concentration of 30 mg NO_3 L^{-1} (EA, 2007a) and 52% of the total river length exceeds phosphorus concentrations of 0.1 mg L^{-1} (EA 2007b).

With aim of protecting and enhancing aquatic ecosystems, the Water Framework Directive (WFD) (2000/60/EC) was introduced in 2000. All water bodies in European Union member states are required to reach "good" and non-deteriorating status by 2015, akin to the conditions observed under minimal anthropogenic influence. Surface waters must achieve good ecological and chemical status, while groundwater must reach a good chemical standard and pose no risk to the status of surface water into which they may flow.

According to Defra (2007) in England and Wales the agricultural practices contribute 70 and 28% of annual nitrogen and phosphorus loads, respectively, figures typical of those reported throughout Europe (IFEN 1997; Torrecilla et al. 2005). The NO_3–N loss is attributed to many factors, including over-fertilization (Lord and Mitchell 1998), excessive manure applications, a failure to consider the nutrient content of manure in fertiliser recommendations, poorly timed nutrient applications, autumn ploughing and intensive stocking of pasture (Shepherd et al. 2001; Shepherd and Chambers 2007).

Smith et al. (1998) reported that phosphorus is lost as sediment bound or particulate phosphorus and in dissolved forms associated with the erosion of phosphorus-enriched soils caused by excess manure/fertilizer applications. The interaction of management practices with the inherent variation in crop, soil type, climate, topography, and hydrology gives rise to large spatial and temporal variation in nutrient concentrations in land runoff. However difficult, diffuse agricultural pollution must be controlled in order to minimise the adverse impacts of agriculture on ecosystems and to comply with legislative requirements. Table 2 shows mitigation measures in controlling the N and phosphorus by agricultural practices according to Cuttle et al. (2007).

2.1 Water Productivity in Rainfed Systems: Mediterranean Region

Rainfed agriculture produces the bulk of the world's food, being practiced in 80% of the world physical agricultural area and generates 62% of the world's staple food (FAO 2008).

Table 2 Mitigation methods for avoiding the nitrogen and P pollution risk under agricultural practices (Cuttle et al. 2007)

Category	Mitigation measures
Land use and soil management	Convert arable land to extensive grassland
	Establish cover crops in the autumn
	Cultivate land for crop establishment in spring rather than autumn
	Adopt minimal cultivation systems
	Cultivate compacted tillage soils
	Cultivate and drill across the slope
	Leave autumn seedbeds rough
	Avoid tramlines over winter
	Establish in-field grass buffer strips
	Loosen compacted soil layers in grassland fields
	Maintain and enhance soil organic-matter levels
	Allow field drainage systems to deteriorate
Livestock management	Reduce overall stocking rates on livestock farms
	Reduce the length of the grazing day or grazing season
	Reduce field stocking rates when soils are wet
	Move feed and water troughs at regular intervals
	Reduce dietary N and P intakes
	Adopt phase feeding of livestock
Fertilizer management	Use a fertilizer-recommendation system
	Integrate fertilizer and manure nutrient supply
	Reduce fertilizer-application rates
	Do not apply P fertilizers to high Pindex soils
	Do not apply fertilizer to high-risk areas
	Avoid spreading fertilizer on fields at high-risk times
Manure management	Increase the capacity of farm manure (slurry) stores
	Minimise the volume of dirty water produced
	Adopt batch storage of slurry
	Adopt batch storage of solid manure
	Compost solid manure
	Change from slurry to a solid-manure-handling system
	Site solid manure heaps away from watercourses and field drains
	Site solid manure heaps on concrete and collect the effluent
	Do not apply manure to high-risk areas
	Do not spread farmyard manure to fields at high-risk times
	Do not spread slurry or poultry manure on fields at high-risk times
	Incorporate manure into the soil
	Transport manure to neighbouring farms
	Incinerate poultry litter
Farm infrastructure	Fence off rivers and streams from livestock
	Construct bridges for livestock crossing rivers and streams
	Re-site gateways away from high-risk areas
	Establish new hedges
	Establish riparian buffer strips
	Establish and maintain artificial (constructed) wetlands

According to FAO (2008) projections food demand in 2030 is expected to be 55% higher than 1998. In response to this demand, global food production should increase at an annual rate of 1.4%.

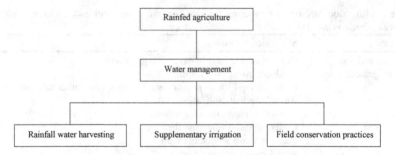

Fig. 2 Water management for rainfed agriculture

Currently, 55% of the gross value of food is produced under rainfed conditions on nearly 72% of the world's harvested cropland. The subject receiving intense debate is the future food demands whether it will be provided by rainfed or irrigated agriculture.

However, what should be clearly understood is that most of the world's food production does not rely on freshwater withdrawals at all and does not necessarily accelerate the naturally occurring rates of evapotranspiration. This means that the bulk of the global agriculture production in the world is rainfed. At the worldwide level, it is well recognised that the potential of rainfed agriculture is large enough to meet present and future food demand through increased productivity (De Fraiture and Wichelns 2010). In this context, an important option is to upgrade rainfed agriculture through better water-, soil-, and land-management practices (Fig. 2). This can be achieved in several ways, including the following options: (1) increasing productivity in rainfed areas through enhanced management of soil moisture and supplemental irrigation where small-scale water storage is feasible; (2) improving soil-fertility management, including the reversal of land degradation; and (3) expanding cropped areas. However, upgrading rainfed agriculture to meet the increasing demand on food is not an easy process but a rather complex one (Rockström et al. 2004). Changes are needed in land, water, and crop management under rainfed agriculture. However, to support these changes, investments are required to build knowledge and to reform and develop institutions. A combination of investments, policy, and research approaches will be needed.

For upgrading rainfed agriculture, we need to consider two major water realities. Firstly, the rainfed agriculture will continue to produce the bulk of the world's food and, secondly, water productivity, which is very low in rainfed agriculture, provides significant opportunities for producing more food with less freshwater. Both realities are strongly connected to each other; hence, increasing the crop:water productivity in rainfed agriculture is the only effective pathway towards attaining food security. A key to success is to invest in the often untapped potential of upgrading rainfed agriculture through integrated water investments. The key challenge is to reduce water-related risks posed by high rainfall variability rather than coping with an absolute lack of water. In rainfed farming, there is generally enough rainfall to double and, even, to quadruple yields but such rainfall is available at the wrong time, causing dry spells and much of the moisture is lost. The temporal and spatial variability of climate, especially rainfall, is a major constraint to yield improvements, competitiveness, and commercialisation of rainfed crops. This is why investment in soil, crop and water management is crucial for upgrading rainfed agriculture (Day et al. 1992; Rockström et al. 2004, 2010).

Evidence from water balance analyses on farmers' fields around the world shows that only a small fraction of rainfall, <30%, is used as productive green-water flow (plant transpiration) supporting plant growth (Rockstrom 2003). The data presented show that rainfall losses through drainage, surface runoff, and non-productive evaporation is extremely high (70 up to 85%), whereas the part of the rainfall used productively, to produce food registers minimum values, between 15 and 30%. According to Oweis and Hachum (2001), in arid areas, only about 10% of the rainfall is consumed as productive green-water flow, with most of the remainder going to non-productive evaporation flow.

Those prevailing conditions imply that investments should be directed to improving rainwater management, which otherwise generates runoff, causing soil erosion and poor yields due to a shortage of soil moisture. Investments in this area will not only maximize rainfall infiltration and the water-holding capacity of the soils, but, in the meantime, will minimize land degradation as well as increasing the water available in the soil.

Water management to upgrade rainfed agriculture encompasses a wide spectrum from water-conservation practices for improving rainwater management on the farmers' fields to managing runoff water (surface and sub-surfaces) for supplying supplemental irrigation water in rainfed food production (Fox and Rockström 2003; Hamdy et al. 2005; Oweis et al. 2005). For instance, studies of rainfed cereal potential under different climate-change scenarios estimated losses at 10–20% of production area with some 1–3 billion people possibly affected in 2080 (Fischer et al. 2005).

Critchley and Siegert (1991) reported several rainwater-management strategies to improve crop yields and green-water productivity. One of these strategies is aiming at maximizing plant-water availability in the root zone. These rainwater-management strategies are dealing with water harvesting and evaporation management:

1. *Water-harvesting strategies: external water-harvesting systems.*
   This strategy involves several management options using several tools, including surface micro-dams, subsurface tanks, farm ponds, percolation dams and tanks as well as diversion and recharging structures. Harvested water can be used for upgrading rainfed agriculture through mitigating dry spells, recharging groundwater, enabling off season irrigation and facilitating multiple uses of water.
2. *In situ water-harvesting systems, plus soil and water conservation.*
   This strategy aims basically at concentrating rainfall water by trapping the surface runoff in cropped area, thereby maximizing infiltration into the soil

matrix. There are several management options which are technically simple and economically sound and, therefore, they are widely used by farmers in many arid and semi-arid regions to capture and reduce rainfall losses (Durán et al. 2006, 2009a, b).

For instance, to maximize rainfall infiltration by reducing runoff can be effectively achieved through terracing, contour cultivation, conservation agriculture, and staggered trenches (Durán et al. 2009a). On the other hand, bunds, ridges, micro-strips, broad-beds and furrows are the management options to capture and concentrate rainfall. Therefore, soil and water conservation or in situ water-harvesting systems should be considered the logical entry-point for improved water management in rainfed agriculture (Wani et al. 2003, Kahinda et al. 2007; Rockström et al. 2010). In many arid and semi-arid regions worldwide, where rainfed agriculture is the dominant producer for cereal crops, highlighted that investments in this type of agriculture have good payoffs in yield improvements and environmental sustainability.

The upgrading and investing in rainfed agriculture has several constraints, including technical, socio-economic, and policy factors and, above all, the inadequate investments in knowledge sharing and scaling-up of best practices. The integrated approach to rainwater management must address links between investments and risk reduction as well as between rainwater management and land and crop management (Trisorio and Hamdy 2008).

The Mediterranean region includes the countries of two groups: the countries of the north coast (from Spain to Turkey) and those of the southern coast (from Morocco to Syria). The climate of the Mediterranean region is characterized by a hot, dry summers lasting from 2 to 7 months, depending on the geographical position from the North to South (Shahin 1996). Winter is mostly rainy, while fall and spring are partially wet (Turner 2004a). Often supplemental irrigation is required by the crops for regular yield and for reducing the inter-annual variability (Oweis 1997). Reasonable yields can be obtained by summer crops if they are irrigated during the entire vegetative cycle (De Boer 1993). In addition, the definition of Mediterranean climate is extended also to other regions of the planet (Estienne and Godard 1970). Southern Australia, South Africa (Cape province), Southern California, and Chile are characterized by Mediterranean-type climates.

The water use in the Mediterranean region shows that about 72% of the available water is used for agricultural sector (Hamdy and Lacirignola 1999). According to Margat and Vallée (1997) water resources are becoming strained, mainly in southern countries and yet often wasted. Farmers are allocated large amounts of water, exceeding crop requirements for all winter and summer crops (Shideed et al. 2005). Their crops are over-irrigated by 30–49% (Hamdy and Katerji 2006). Attaining the UN Millennium Development Goals to halve the number of poor and food insecure by 2015 poses a tremendous development challenge. Until recently the focus on agricultural water policy has centred around withdrawals, allocation, and management of runoff-water flows—that is, surface flow in rivers and subsurface groundwater flow, which are defined as blue water flows, and the subsequent blue-water runoff regenerate aquifers, lakes, wetlands, and constructed water-storage facilities. The blue-water domain concerns the global water crisis, due to growing per capita water scarcity (30% of the world's population projected to face water stress by 2030) and over-expropriation of blue-water resources, primarily for irrigation, resulting in drying rivers, falling groundwater levels, and declining lake and wetland systems (Rosegrant et al. 2002). On the other hand, as stated by Falkenmark and Rockström (2004) green water sustains rainfed agriculture, which is practiced on 80% of the world's agricultural land, and generates 60–70% of the world's food.

The sustainable use of water resources in the European Mediterranean basin is crucial, as is the solution of the many problems generated by water scarcity and misuse, particularly in the southern and eastern parts of the region, (Zacharias and Koussouris 2000; Mariolakos 2007; Downward and Taylor 2007). The high irrigation needs and changes in consumer demands (especially after population shifts from rural to urban areas and because of increasing tourism and industrialisation) are the main problems in the region. The proper management of limited water resources in the Mediterranean involves several research areas, most of which are directly related to agriculture, concerning the improvement of water and nutrient use in agriculture through the management and breeding of irrigated and rain-fed crops. However, these fields of research address only one side of a complex problem that challenges water sustainability in the region (Araus 2004).

The main cause of current environmental problems in the Mediterranean basin is human-made; large-scale soil erosion, pollution, and food shortage. Depletion of non-renewable groundwater is widespread, and remaining water resources are often polluted (Sánchez et al. 2003; Yang et al. 2006). Salt–water intrusion is common in many of the coastal aquifers. Most of the agricultural production in the Mediterranean basin comes from irrigated areas (Zhang and Oweis 1999; Alexandridis et al. 2008; Wriedt et al. 2009). Hence agriculture is the main consumer of water in the region with 80% on average (FAO 2003). Water for irrigation is even scarcer than land, and it is becoming increasingly harder to find land suitable for irrigation. The irrigated sector will have to face major challenges with the new scenario of free agricultural trade: the food strategy may change with the possibility of some products supplied by the world market; a part of the water resources may be reallocated to high added-value exports instead of basic production or to industrial activities, tourism, and domestic water supply (Postel 1999; Bahri 2002).

In a mountainside within a Mediterranean watershed with rainfed olive (*Olea europaea*) trees under different soil management systems (non-tillage with barley strips of 4 m width; non-tillage with native vegetation strips of 4 m width; and non-tillage without plant strips) were monitored for soil–water content (SWC) with multi sensor capacitance probes at two positions (beneath the tree and in the middle of plant strips). The soil water contents beneath the trees (SWC-bt) in all treatments were higher than those of the plant strip positions (SWC-st) except in the non-tillage without plant strip treatment, where the mean SWC-bt at most depths was almost the same as SWC-st, with the exception of 10-cm depth, which had a SWC-bt higher (0.07 cm^3 cm^{-3}) than SWC-st (0.05 cm^3 cm^{-3}) during the monitoring period. The mean SWC-bt for the entire 100-cm soil profile for the non-tillage with barley strips, non-tillage with native-vegetation strips, and non-tillage without plant-strip treatments was 0.13, 0.11 and 0.09 cm^3 cm^{-3}, respectively. During each season, the mean SWC-bt was higher than the mean SWC-st at all depths for the non-tillage with barley strip treatment, at 10, 20, and 30-cm depths for the non-tillage with native vegetation strip treatment, and only at 10-cm depth for the non-tillage without the plant-strip treatment (Fig. 3). The increased recharge of soil water of the plant-strip positions is attributable to canopy interception and subsequent stem flow. Also, plant strips slow down runoff and encourage infiltration. Under these experimental conditions (30% slope) with a soil of relatively low permeability, the subsoil could restrict deep percolation of infiltrating precipitation and induce lateral subsurface flow. Therefore, most soil water could be available for the root system of olive trees, which is the next in the gravitational trajectory of both the surface and subsurface flow of water, and thus the plant strips could act as sinks for rainfall and overland flow.

Around the Mediterranean basin, the degradation of soil and water resources is a serious threat for the human welfare and the natural environment as a result of the unique climate, topography, soil characteristics and peculiarities of agriculture.

The negative impacts on water resources include pollution due to nutrient and pesticide leaching together with seawater intrusion into aquifers. Moreover, the dramatic change of agricultural practices during the last 50 years is one of the main driving forces for environmental degradation in this region, especially through its impact on soil and water resources. Although these changes have had many positive effects on farming, there have also been significant costs. In the last decades there is increasing interest in crop-production systems that optimise yields while conserving soil, water, and energy at the same time as protecting the environment (Stamatiadis et al. 1996).

Agriculture has direct as well as indirect effects on the quality of soil and water resources. These effects depend both on macro-scale factors (e.g. climate, landscape topography and parent material) as well as micro-scale factors (e.g. management and the land use at a watershed or a farm scale). These factors characterize the ecosystems based on the similarity of inputs (Odum 1983), and establish the type of certain agricultural practices that are possible. Because of the unique site characteristics and the agricultural peculiarities of the Mediterranean, the main impact of agriculture on soil quality is erosion, salinization, compaction, reduction of organic matter, and non-point source pollution. As a consequence, the soil degradation lowers water quality through leaching of pesticides and excess nutrients into surface and ground water together with seawater intrusion into aquifers. According to Maltby (1991), the scientific basis is still incomplete for explaining how different ecosystems work and how different environmental factors interact to control functioning.

Soil degradation affects about a 1/5 the arid domain in the Mediterranean basin, but mainly the semi-arid margins, which are generally cultivated. Degradation of the soil and vegetation cover can lead to desert landscapes (Thirgood 1981;

Fig. 3 Mean soil–water contents of different positions at various depths during each season for the different soil management systems: non-tillage without plant strips (NT); non-tillage with barley strips (BS); and non-tillage with native-vegetation strips (NVS)

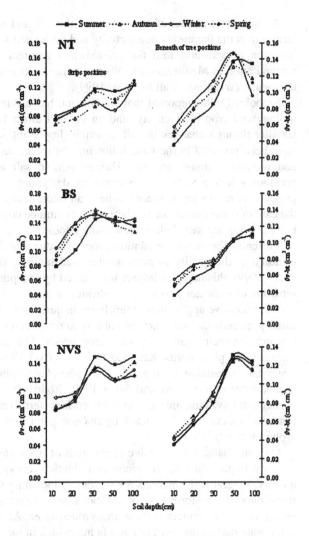

Rapp 1987; Thomas and Middleton 1994a, b; Durán et al. 2006). This desertification is a consequence not only of climatic change but also human activities (vegetation removal, wood exploitation for fuel, and over-grazing). Land degradation in the Mediterranean area has increased for a variety of reasons and is estimated to threaten over 60% of the land, especially in southern Europe (UNEP 1991). The renewed pressure on land resources through migration, the changes in agriculture both in terms of what is produced (cash-crops) and the mode of production (intensive agriculture), the increased demands of water through the development of irrigation schemes, in addition to the impact of land degradation on flooding, ground water recharge, saltwater intrusion and soil salinization has in many cases been responsible for land desertification (CEC 1994).

An important impact of agriculture on the soil and water quality in the Mediterranean is the increasing conductivity and associated salinization of soils and the intrusion of seawater into the groundwater aquifers near the coast. The salt accumulation in Mediterranean soils is a natural process favoured by the region's ecological conditions (Zalidis et al. 1999).

Szabolcs (1996) reported that salinization has a direct negative effect on soil biology and crop productivity, and an indirect effect leading to the loss of soil stability through changes in soil structure. Important areas of salt-affected soils have been revealed in the past, following the numerous large-scale flood control and wetland drainage projects. Human-induced salt accumulation occurred in previous salt-free soils due to errors in designing and constructing irrigation projects. In recent years, there has been an effort to correct these errors except in the case of some coastal areas where the low altitude constrains the maintenance of naturally trapped salts below the rooting zone.

Currently the salt-accumulation process is due mainly to the continuing deterioration of the quality of groundwater used for irrigation. According to Zalidis et al. (1999), this negative impact was caused by overpumping and the consequent intrusion of seawater into the groundwater strata.

The excessive application of fertilizers in quantity and frequency by agriculture usually exceeds the soil functional ability to retain and transform the nutrients and synchronize their availability with crop needs. The saturation of the soil with nitrogen and phosphorus, have led to nitrate (NO_3) losses into shallow groundwater and saturation of the soil with phosphate, which may also move into groundwater (Breeuwsma and Silva 1992; Rodriguez et al. 2008). In intensive horticultural systems, interaction between high fertilizer inputs and major irrigation schemes enhances NO_3 leaching and non-point source pollution of surface and ground water (EEA 1995).

On other hand, the extensive application of pesticides in agricultural land has negative impact on both the biotic and abiotic processes within the soil. Consequently, several soil functions are degraded, including the food-web support, the retention and transformation of toxicants and nutrients, soil resilience, and the ability of soil to protect surface and groundwater. At the farm scale, pesticides deteriorate part of the soil flora and fauna, which in turn causes both physical and chemical deterioration (Doran et al. 1996; Müller et al. 2007). At the watershed scale, the main problem derives from the leaching and drainage of pesticides into the surface and ground water (Domagalski and Dubrovsky 1992; Van der Zee and Boesten 1993; Hantush et al. 2000). In addition, the reduction of the soil's ability to remove other pollutants, mainly due to the alteration of soil properties and the degradation of soil's toxicant retention and transformation function, enhances the transport of these pollutants to adjacent water bodies.

The major physical impact of agricultural practices on Mediterranean soils involves compaction and water erosion (Hamza and Anderson 2005; Lagacherie et al. 2006; Koulouri and Giourga 2007; Blavet et al. 2009). Soil compaction is caused by the repetitive and cumulative effect of heavy machinery (Van Dijck and Van Asch 2002).

The resulting decrease of soil porosity reduces root penetration and access to the soil nutrients and alters biological activity on the farm scale. On the watershed scale, soil compaction increases surface runoff since less rainwater is able to percolate. This increases the risk of water erosion, loss of topsoil and nutrients, and non-point source pollution of water resources (EEA 1995).

The negative effects of erosion include diminished infiltration rates and water availability, loss of organic matter and nutrients and an ultimate loss of production potential (Hillel 1991). Downstream eventualities include disrupted or lower-quality water supplies, silting that impair drainage and maintenance of river channels and irrigation systems, and increased frequency and severity of floods (Pimentel et al. 1995).

Loss of soil organic matter reduces root penetration, soil moisture and permeability, which in turn increases the risk of erosion and surface runoff and reduces biological activity of soils (EEA 1995). The microbial functions help maintain a soil system with available nutrients, aiding stability and thereby reducing erosion and augmenting water-holding capacity (Kennedy and Gewin 1997).

Among the diverse water users in the region, the agricultural sector shows the highest water losses. The low irrigation efficiency can be attributed mainly to water mismanagement and also technical problems of conveyance, distribution, and on-farm application, plus poor maintenance of irrigation structures, often because of inadequate resources for operation and maintenance.

Irrigation agriculture in the Mediterranean basin is usually governed by the priority of quantity versus quality of the irrigation systems and prioritisation of large rather than small irrigation networks. Although the high priority and large investments have been directed at developing water resources to meet irrigation needs, the performance of large public irrigation systems has fallen short of expectations. Crop yield and efficiency in water use are lower than originally projected (Postel 1999).

The improvement of rainfed agriculture also has an important water- and food-security relevance in this region. Firstly, most of the food for populations in rural areas depends on this type of agriculture. Secondly, most rainfed lands have low productivity. Consequently, an increase in production would reduce the growing water demand in agriculture.

The most important natural resource in the drier environments is rainfall and despite its scarcity, rainfall is generally poorly managed and much of it is lost through runoff and evaporation. The collection and efficient use of rainwater is fundamental.

In addition, different technologies and management strategies are available to increase water-use efficiency and sustainability of agriculture. Figure 4 shows, according to Mannion 1995b, the complex multiplicity of factors involved in agriculture that may affect its performance and long-term sustainability.

In a broad sense, the water productivity in rainfed areas can be increased in many ways: by promoting water-efficient irrigation (deficit irrigation), water-efficient agriculture (precision agriculture) and water re-use in agriculture, particularly the recycling of drainage water and reuse of urban wastewater.

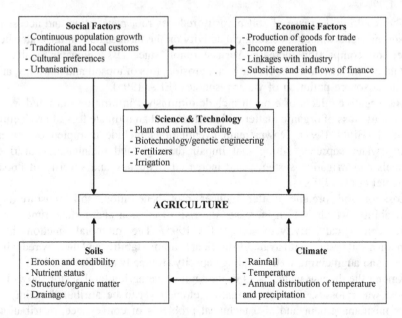

Fig. 4 Factors involved in agricultural performance and sustainability

The management of irrigation water can be achieved by better on-farm water management, reducing distribution losses, changing cropping patterns, improving irrigation systems, and adopting irrigation-efficient technology.

The reuse of urban wastewater and brackish waters as well as the more conventional methods of improved water delivery need to be encouraged. Thus, on-farm irrigation methods, such as sprinklers and drip irrigation (Suarez 1992) can considerably reduce water loss. In addition to their potential for increasing water-use efficiency, these two irrigation systems can be used to farm infertile lands and sandy and rocky soils.

In addition, to increase water productivity, well-managed irrigation schemes can overcome salinization and related problems. Another approach is the breeding and cultivation of either salt-tolerant crops or crops with improved water use and water productivity.

For rainfed agriculture, solutions include the development of drought-resistant crop varieties such as cereals, alternate tillage practices to conserve water, or simple water-harvesting systems to provide access to water at the critical growth stages of the crops, such as planting cereal strips on sloping lands with almond and olive orchards (Francia et al. 2006; Durán et al. 2009a, b). According to Araus et al. (2003), even though breeding for drought tolerance may lower yield potential, it increases yield stability over many seasons, which could still offer long-term benefit to many farmers in drought-prone areas.

According to Rockström and Barron (2007), from a water perspective, there are two main ways to upgrade rainfed agriculture: (1) increase water-uptake capacity

Table 3 Integrated soil and water management for upgrading rainfed agriculture systems

Strategy for upgrading	Management	Methodology	Objective
Plant water uptake capacity	Soil	Tillage	Root length and density
		Crop rotation	Crop development
		Mulching, manures and fertilizers	
	Crop	Crop choice	
		Intercropping/crop rotation	
		Timing operations	
		Pest management	
Plant water availability	Soil	Tillage	Soil infiltrability
		Weed management	Less unproductive competition
		Soil and water conservation	
		Crop rotation	
		Mulching/organic manures	Water-holding capacity
	Water	Water harvesting	Dry spell mitigation

of the crop, and (2) increase water availability for the plant. Even though these strategies focus on water, the approaches and practices to achieve them are not necessarily solely associated to water management (Table 3).

The water-uptake capacity of the plant could be improved through proper crop and soil management, maximizing depth and density of roots and development of canopy and grain (Rockström and Barron 2007). Other agricultural practices such as tillage, crop rotations, mulching and the use of organic manures (green and animal) will improve the soil structure, and consequently root development. Also, crop choice, intercropping, timing of operations, weeds and pest management will influence plant water uptake capacity. Plant breeding and genetic development can improve the harvest index (the ratio of grain to total biomass), water-uptake capacity of the root, and crop resistance to stress from, for example, water and disease. Organic and inorganic soil-fertility management is a prerequisite to crop growth, and therefore the key to the crop's water-uptake capacity. Soil conditions at the surface, i.e. infiltration capacity, and the soil structure in the top soil, will also affect plant water availability. Soil- and water-conservation practices, which focus on maximizing rainfall infiltration will, together with crop rotation, mulching and manure management, affect plant-water availability. Water management, for example through water harvesting practices, can mitigate dry spells, thereby securing plant water over time.

Stewart (1988) reported that the timing of operations is crucial to maximize the crop response to erratic rainfall in semi-arid environments. Early soil preparation and dry planting, which enables a full crop response to the rainfall at the beginning of the rainy season, may constitute the difference between getting a crop and complete crop failure. Such rainfall may account for up to 20% of the total seasonal rainfall.

As agriculture, the main consumer of water in the Mediterranean region is currently faced with the challenge of new approaches to water-resource management that ensures the protection of water resources. This aim can be achieved by the following measures: (i) to save water by controlling the water supply according to the proper determination of crop water requirements and accurate irrigation scheduling based on biological and physical criteria; (ii) to improve the performance of irrigation systems; and (iii) to optimise the water-use efficiency of crops cultivated in the Mediterranean area. These approaches could be developed through the identification of environmental, biological, and agricultural parameters which improve the water-use efficiency of crops.

2.2 Rainwater-Harvesting Practices

Water harvesting and related techniques such as "runoff agriculture" may provide much more water to rainfed crops than natural rainfall would alone. It is not a new concept; indeed in the Mediterranean regions of the world there is much indigenous knowledge about this ancient practice. Besides its technical feasibility, a critical element of success of traditional water-harvesting systems is a social structure and set of community norms and practices to which farmers adhere (Postel 1999). Several studies have addressed the benefits from combining indigenous with modern knowledge and have explored the potential of several techniques for water harvesting and their adaptation to local conditions. Figure 5 shows the water-harvesting methods used by the ICARDA (International Center for Agricultural Research in the Dry Areas).

The rainfall-water harvesting is a promising practice to support sustainable development in many parts of the world facing climate change. The rainfall-water harvesting practices improve hydrological indicators such as infiltration and groundwater recharge (Wakindiki and Ben-Hur 2004; Zougmore et al. 2003)). Soil nutrients are enriched (Schiettecatte et al., 2005). Biomass production increases, with subsequent higher yields (Singh et al. 1998; Ellis and Tengberg 2000; Wakindiki and Ben-Hur 2004; Pretorius et al. 2005). Higher biomass supports a higher number of plants and animals, although native species might be replaced by crops as the landscape might change as a whole (Ludwing et al. 2005). In the context of agricultural production in drylands, soil- and water-conservation practices such as rainfall-water harvesting provide an opportunity to stabilize agricultural landscapes in semiarid regions and to make them more productive and more resilient against climate change (Wallace 2000; Lal 2001; Li 2003). Stabilization of the agricultural landscape includes the restoration of degraded cultivated and/or natural grazing lands. There are many marginal water sources that could be used more efficiently such as road and land runoff, which is normally lost through erosion (Prinz and Malik 2002). Among the most common soil- and water-conservation techniques, rainwater harvesting is massively pursued in many countries (Stroosnijder 2003; Batchelor et al. 2002; Pandey et al. 2003). In addition, rainfall-water harvesting is also one of the practices recommended by UNCCD to combat desertification.

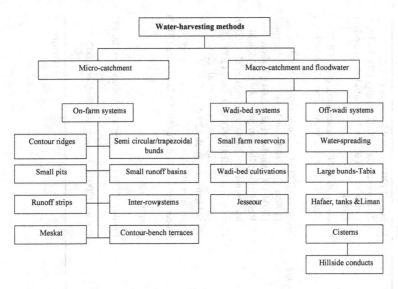

Fig. 5 Classification of water-harvesting methods

Rainfall-water harvesting is generally considered to be beneficial only in this respect but the main problems are low rates of adoption (Nji and Fonteh 2002; Bodnar and de Graaff 2003; Woyessa et al. 2005) or failed adoption processes due to insufficient application (Aberra 2004).

Rainfall-water harvesting improves resilience in enhancing socio-economic as well as ecological adaptability. Socio-economic adaptability may be improved by increased food security, extra income and consequently enhanced and sustainable livelihoods. Also, rainfall-water harvesting practices used to restore degraded areas improve resilience by improving productivity (Lal 2001, 2004; UNCCD 2001; Rockström 2004). Humans therefore are not only a cause of degradation but can be the driving force for restoration processes due to their efforts.

Table 4 shows the impact on landscape functions by rainfall-water harvesting practices in African drylands according to Vohland and Barry (2009).

2.3 Water-Use Efficiency

Water-use efficiency is a broad concept that can be defined in many ways. For farmers and land managers, water-use efficiency is the yield of harvested crop product achieved from the water available to the crop through rainfall, irrigation, and the contribution of soil–water storage (Van Duivenbooden et al. 2000; Tennakoon and Milroy 2003; Pala et al. 2007). The main ways to enhance water-use efficiency in irrigated agriculture are: to increase the output per unit of water

Table 4 Impact of rainfall-water harvesting practices on landscape functions in African drylands

Landscape function	Indicators	Response	Spatial scale	Reference
Ground-water recharge	Level of wells	Improved infiltration	Catchment area	Mohamed (2002); Woldearegay (2002); Zougmore et al. (2003); Ngigi (2003a, b); Botha et al. (2004); Wakindiki and Ben-Hur (2004)
Wet land/aquatic ecosystem maintenance	Plant species and freshwater zoonoeses	Competition on superficial water/ improved infiltration	River, wetland	Oweis et al. (1999); Falkenmark et al. (2001); Ngigi (2003a, b); Woyessa et al. (2005)
Biomass production	Yield	Rainfall water and water use efficiency	Plant-field-landscape	Roose et al. (1993); Tabor (1995); Singh et al. (1998); Ellis and Tengberg (2000); Barron (2004); Kayombo et al. (2004); Kabore and Reij (2004); Rockström (2004); Wakindiki and Ben-Hur (2004); Pretorius et al. (2005)
Nutrient cycling	Soil nutrients	Sediment trapping	Field	Herweg and Ludi (1999); de Graaff (2000) Fatondji (2002); Zougmore et al. (2003); Schiettecatte et al. (2005)
Floral diversity	Biological activity	Attraction e.g. termites	Field	Fatondji et al. (2001)
	Species diversity	Plant growth in bare soils	Field	Ouedraogo and Kabore (1996); Bangoura (2002)
	Endemic/local species	Crop species instead of savanna matrix	Community	Warren (1995)
Animal diversity	More species	Enhanced biomass for food and shelter	Field landscape	Pandey (2001), Herweg and Steiner (2002a, b)
	Higher abundance	Enhanced biomass for food and shelter	Field landscape	Pandey (2001); Herweg and Steiner (2002a,b)

(continued)

Table 4 (continued)

Landscape function	Indicators	Response	Spatial scale	Reference
Food security	Yields in bad years	Bridging of dry spells	Landscape	van Dijk and Ahmed (1993); Ellis and Tengberg (2000); Rockström et al. (2002, 2004)
Water availability for society	Time spent water collecting	Improved groundwater recharge	Community	Falkenmark et al. (2001)
	Community rules	Upstream–downstream competition on water	Community	Ngigi (2003a, b); Rockström et al. (2004)
Income	Money	Enhanced yields	Household	Ellis and Tengberg (2000); Kunze (2000); Zaal (2002)
	Money	Adoption rate	Community	Bangoura (2002)

(engineering and agronomic management aspects), reduce losses of water to unusable sinks, reduce water degradation (environmental aspects), and reallocate water to higher priority uses (societal aspects).

Ameliorating water-use efficiency in agriculture will require a boost in crop-water productivity (greater marketable crop yield per unit of water removed by plant) and a reduction in water losses from the root zone, a critical zone where adequate storage of moisture and nutrients are required for optimising crop production. Consequently, the amount of water required for food production depends on the agricultural commodities produced.

Improving water-use efficiency by 40% on rainfed and irrigated lands could reduce the need for additional withdrawals for irrigation to zero over the next 25 years. However, this is a big challenge for many countries. Increasing water-use efficiency is a paramount objective, particularly in arid and semi-arid areas with erratic rainfall patterns. Under rainfed conditions, soil water can be lost from the soil surface through evaporation or through plant uptake and subsequently lost via plant transpiration (Bennie and Hensley 2001; Meerkerk et al. 2008). It can also be lost through runoff and deep infiltration through the soil, as it is essential to apply appropriate measures for its control (Durán et al. 2009a, b; Thierfelder and Wall 2009).

When irrigation is available, water losses also include the mismanagement of irrigation water from its source to the crop roots. Usually, more than 50% of irrigation water is "lost" for the crop at the farm level. However, at the watershed level, it might be less due to possible recoveries from the subsoil and groundwater. These off-site losses of water can result from either inappropriate land-management practices to capture a substantial part of the rainfall within an agricultural landscape and retain it in the root zone or excessive use of irrigation water.

Such losses lead not only to water waste but also potential hazards of soil salinity and water pollution resulting from the transport of nitrate, phosphate, sediments, and agro-chemicals to streams, lakes and rivers.

Many promising strategies for raising water-use efficiency include appropriate integrated land–water management practices as follows: (1) adequate soil fertility to remove nutrient constraints on crop production for every drop of water available through either rainfall or irrigation; (2) efficient recycling of agricultural waste-water; (3) soil–water conservation measures through crop residue incorporation, adequate land preparation for crop establishment, and rainwater harvesting; and (3) conservation tillage to increase water infiltration, reduce runoff and improve soil-moisture storage. In addition, novel irrigation technologies such as supplementary irrigation (some irrigation inputs to supplement inadequate rainfall), deficit irrigation (eliminating irrigation at times that have little impact on yield) and drip irrigation (targeting irrigation water to root zones) can also minimize soil evaporation and provide nutrients by fertigation, thus making more water available for plant transpiration.

One of the components of a management system that affects water-use efficiency is soil fertility (Fan et al. 2005; Aboudrare et al. 2006; Cayci et al. 2009). A complete and balanced fertility program helps to produce a crop with roots that explore more soil volume for water and nutrients in less time. This results in a healthier crop that can more easily withstand seasonal stress. Recent research in an area where water is a major concern has demonstrated the importance of balanced fertility and irrigation methods in maximizing water-use efficiency (Raun and Johnson 1999; Mohammed et al. 1999; Tayel et al. 2006).

Improvement of water-use efficiency of field crops is an imperative imposed by the critical situation of water resources. As expected, a large range of water-use-efficiency values could be observed, for the same species, which can be ascribed mainly to: (1) fertilizers and water management (e.g. water regime, mineral supply, and water quality); (2) plant factors (e.g. species, cultivar, and sensitivity of growth stage to the stress); and finally (3) environmental factors (e.g. climate, atmospheric pollution, soil texture, and climate change).

2.3.1 Estimating the Water-Use Efficiency

There are two approaches for considering water-use efficiency. On the one hand, there is the eco-physiological approach, which is based on the analysis of the relationship between photosynthesis and leaf transpiration per unit area, at the leaf scale (Pearcy 1983), canopy scale (Steduto et al. 1997), and territorial scale (Chen and Coughenour 2004). This approach leads to the following results: (1) description of the processes determining theoretical water-use efficiency (Cowan 1982; Farquhar and Sharkey 1982; Hsiao 1993) and (2) comparison of leaf photosynthesis and transpiration capacity of a species cultivated under different irrigation regimes and to analyse the theoretic consequences on water-use efficiency (Cheesman 1991; Leuning 1995; Katerji and Bethenod 1997).

On the other hand, the agricultural approach is based on water consumption and yield, describing water-use efficiency on various scales from the leaf to the field. In its simplest terms, this approach is characterized as crop yield per unit of water use (Sinclair et al. 1984). The time scale considered is the whole vegetative cycle, this providing crucial data to manage irrigated crops and thereby improve yield.

The water-use efficiency (WUE) can be calculated by following equation:

$$\text{WUE} \left(\text{kg m}^{-3}\right) = \text{yield/water consumption} \tag{1}$$

The yield parameter (kg m^{-2}) could be indicated by (1) global dry-matter yield or (2) marketable crop yield. However, the latter is a more useful criterion than dry matter because of the large variety of crops (biomass, grains, fruits, oil, etc.). Moreover, marketable yield is more precise because it represents economical value, which is a basic factor in determining irrigation cost.

Of the water used by crops during the growing season, 99% is released as water vapour into the atmosphere, and the crop water being used is considered approximately equal to evapotranspiration (ET). Soil evaporation is usually <10% of the total ET, even when the soil surface remains quite wet (Jara et al. 1998; Villalobos and Fereres 1990). The correct management of different agro-techniques (e.g., sowing time, plant density, irrigation frequency, fertilisation, use of mulching) can minimize the evaporation fraction of total evapotranspiration (Turner 2004b; Deng et al. 2006). Brown (1999) pointed out that the upcoming benchmark for expressing yield may be the amount of water required to produce a unit of crop yield, which is simply the long-used transpiration ratio, or the inverse of water-use efficiency. Bos (1980, 1985) recommended that water-use efficiency for irrigation be based on the yield produced above the rainfed or dryland yield divided by the net ET difference for the irrigated crop, which he called the yield:ET ratio. He also proposed the irrigated difference from the dryland yield divided by the gross applied water, which he called the yield:water-supply ratio and is referred to as irrigation-water-use efficiency.

Defining water-use efficiency for irrigation is additionally complex because the scale of importance for the water resource shifts to the broader hydrologic, watershed, irrigation district, or irrigation project scale, and the water components may not be so precisely defined, becoming even more qualitative when such terms as reasonable, beneficial, or recoverable are used (Burt et al. 1997). Irrigation can be an effective means of improving water-use efficiency through increasing crop yield, especially in semiarid and arid environments. Even in wet environments, irrigation is particularly effective in overcoming short-duration droughts.

According to Katerji and Perrier (1985), complex models are necessary to determine the portion of soil evaporation in ET, and therefore scientists determine the ET in order to evaluate water-use efficiency.

At the plot scale, ET can be estimated through different approaches: (1) ET directly measured by using drainage/weighing lysimeters or indirectly through micrometeorological methods (Bowen ratio, aerodynamic), (2) ET calculated through the soil–water balance, (3) ET estimated according to the FAO method (Allen et al. 1998), and (4) ET simulated through different productivity models.

The overestimation of water supplied to crops is one of the characteristics of irrigation practice making it difficult to understand the calculated water-use efficiency values (Shideed et al. 2005).

Water is important in rainfed agriculture, critically important in semiarid dryland agriculture, and explicitly important in irrigated agriculture. Wallace and Batchelor (1997) offered four options for enhancing water-use efficiency in irrigated agriculture (Table 5). They point out that focusing on only one category will likely be unsuccessful. Bos (1980, 1985) developed expressions that can, perhaps, more consistently discriminate the role that irrigation has in WUE. His expressions can be written for ET_{WUE} and I_{WUE} as:

$$ET_{WUE} = (Yi - Yd) / (ETi - ETd), \tag{2}$$

$$I_{WUE} = (Yi - Yd) / Ii, \tag{3}$$

where Yi is the yield and ETi is the ET for irrigation level i, Yd is the yield and ETd is the ET for an equivalent dryland or rainfed only plot, and Ii is the amount of irrigation applied for irrigation level i.

Table 5 Available options for improving irrigation efficiency at a field level (Wallace and Batchelor, 1997)

Category	Options
Agronomic	Crop management to enhance rainfall-water harvesting or reduce water evaporation (i.e. crop debris, conservation till, and plant spacing); improved varieties; advanced cropping strategies that maximize cropped area during periods of lower water demands and/or periods when rainfall may have a greater likelihood of occurrence
Engineering	Irrigation systems that reduce application losses, improve distribution uniformity, or both; cropping systems that can enhance rainfall harvesting, e.g. crop debris, deep chiselling or paratilling, furrow dyking, and dammer-dyker pitting
Management	Demand-based irrigation scheduling; slight to moderate deficit irrigation to promote deeper soil–water extraction; avoiding root-zone salinity yield thresholds; preventive equipment maintenance to reduce unexpected equipment failures
Institutional	User participation in irrigation-district operation and maintenance; water pricing and legal incentives to reduce water use and penalties for inefficient use; training and educational opportunities for learning, advanced techniques

In most arid areas, Yd would be zero or small; however, ETd could be much greater than zero and variable, depending on agronomic practices. In semiarid and rainfed areas, Yd could be determined several ways. In the strictest sense, it would be the yield under exactly the same management as the i treatment or system but without irrigation. In a more comparative system, it might be estimated by yields from comparable dryland or rainfed plots that were not irrigated. Often, however, agronomic practices differ substantially between dryland and/or rainfed and irrigated practices (e.g., variety, sowing date, fertility management, pest management, sowing density, and planting geometry). Thus, results that are quite different might be found for Yd and ETd based on differences in management. The water use in Eq. 1 is difficult to determine with precision. Thus, in some situations, benchmark water-use efficiency (WUE_b) is used by many irrigation practitioners. It can be defined as:

$$WUE_b = Y(\text{usually the economic yield})/(Pe + I + SW), \qquad (4)$$

where Pe is effective rainfall, I is irrigation applied, and SW is soil–water depletion from root zone during the growing season. The denominator of Eq. 4 is a surrogate estimate for the water used to produce the crop, depending on the neglect of percolation, groundwater use, and surface runoff.

According to Howell (2001), many agronomic, engineering, and management technologies can reduce non-productive water use in irrigated agriculture. However, in some cases, increasing irrigation efficiency may not simply gain new water for allocation unless the consumptive use of the diverted water is actually reduced. In this context, Seckler (1996) summarized these opportunities as (i) increasing the output per unit of ET (essentially water-use efficiency), (ii) reducing losses of usable water to sinks, (iii) reducing water pollution (from sediments, salinity, nutrients, and other agrochemicals), and (iv) reallocating water from lower-valued

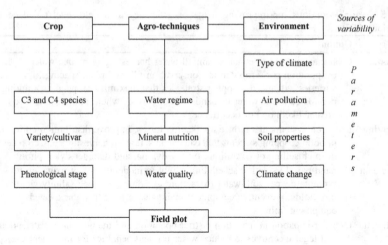

Fig. 6 Sources of variability and parameters involved for water-use efficiency determination

to higher-valued uses. The latter opportunity can be positive or negative to agriculture, depending on how secondary and tertiary interest holders are addressed.

Katerji et al. (2008) excluding experimental errors related to the determination of yield and ET, the variability in determining water-use efficiency can be ascribed to mainly three sources: (1) agro-techniques: water and fertilizer applied to crops and analysed in terms of quantity and quality; (2) plant: differences between species, variety effects, phenological stage, sensitivity to water constraints; (3) environment: soil and climate. These factors include evaporative demand, atmospheric pollution and climatic changes (Fig. 6).

The schematic overview described in Fig. 6 underlines the parameters susceptible to interference in the determination of water-use efficiency of a species at the field-plot scale. Katerji et al. (2008) pointed out the lack of the current knowledge in order to identify the promising future of this field of study, considered from two standpoints: methodology and research.

For the methodological level, particular attention needs to be given to the methods of ET determination because the absence of a correct estimation of this parameter should be lack of a reliable water-use-efficiency analysis.

Correct analysis water-use-efficiency data requires reliable complementarity between the eco-physiological and agricultural approaches. Concretely, the methodology that associates an indicator of plant-water status with soil–water-status criteria in analysing water-use efficiency can reduce the variability observed for water-use efficiency through standard experimental conditions (Li et al. 2000; Wang et al. 2007; Ritchie and Basso 2008).

The reduction of water-use-efficiency variability is a primary condition for appropriate analysis of water-use efficiency. This is so for the designated research to examine the well-founded management methods in order to improve water-use efficiency (modification of sowing dates, mulching).

Identification of the phenological stages quantitatively and qualitatively sensitive to water constraint under field conditions is preliminary for rational complementary irrigation practice.

Research on water-use efficiency should be developed for many crops by determining water consumption. The current research in this sense is conducted with high water quality (Hassanli et al. 2010). Since saline soils are increasing, due to steadily more frequent use of non-conventional waters for irrigation, the study of water-use efficiency in relation to water quality becomes crucial, especially with saline waters (Lea and Syvertsen 1993; Smith et al. 2010; Grewal 2010). Plant amelioration has made important progress through the creation of varieties with higher drought and salinity resistance.

Research relating water-use efficiency to the mineral supply is limited, as it is necessary to demonstrate to farmers the importance of well-founded fertilizing practices.

The analysis of the relationship between water use efficiency and air pollution is still not developed.

Climate change could lead to major transformations of agricultural practices in the future (species and variety choice, sowing and yielding date, irrigation practice, water-use efficiency), and therefore it is necessary to prepare agriculture to face these imminent future changes.

The improvement of water-use efficiency especially en arid and semiarid regions is an imperative imposed by the critical situation of water resources as well as by the demographical surge.

3 Water Resources and Sustainable Agriculture

Of the estimated $1.4 \times 10^{18} \text{m}^3$ of water on the earth, more than 97% is in the oceans (Shiklomanov and Rodda 2003). Approximately $35 \times 10^{15} \text{ m}^3$ of the Earth's water is freshwater, of which about 0.3% is held in rivers, lakes, and reservoirs. The remainder of freshwater is stored in glaciers, permanent snow, and groundwater aquifers. The earth's atmosphere contains about $13 \times 10^{12} \text{ m}^3$ of water, and is the source of all the rain that falls on earth (Shiklomanov and Rodda 2003). Yearly, about 151,000 quads (quad $= 10^{15}$ BTU British Thermal Unit) of solar energy cause evaporation and move about $577 \times 10^{12} \text{ m}^3$ of water from the earth's surface into the atmosphere. Of this evaporation, 86% is from oceans (Shiklomanov 1993). Although only 14% of the water evaporation is from land, about 20% ($115 \times 10^{12} \text{ m}^3 \text{ yr}^{-1}$) of the world's precipitation falls on land with the surplus water returning to the oceans via rivers (Shiklomanov 1993). Thus, each year, solar energy transfers a significant portion of water from oceans to land areas. According to Jackson et al. (2002) this aspect of the hydrologic cycle is vital not only to agriculture but also to human life and natural ecosystems.

Although water is considered a renewable resource because it depends on rainfall, its availability is finite in terms of the amount available per unit time in any one region. The average precipitation for most continents is about 700 mm yr^{-1} (7 million L ha^{-1} yr^{-1}), but varies among and within them (Shiklomanov and Rodda 2003). In general, a region is considered water scarce when the availability of water drops below 1,000,000 L $capita^{-1}$ yr^{-1} (Engleman and Le Roy 1993). Africa, despite having an average of 640 mm yr^{-1} of rainfall, is relatively arid since its high temperatures and winds that cause rapid evaporation (Vorosmarty et al. 2000; Ashton 2002). Regions that receive low rainfall (<500 mm yr^{-1}), undergo serious water shortages and inadequate crop yields. For example, 9 of the 14 Middle Eastern countries (including Egypt, Jordan, Israel, Syria, Iraq, Iran, and Saudi Arabia) have insufficient rainfall (Myers and Kent 2001; UNEP 2003).

In terms of water availability, about 30% (11×10^{15} m^3) of all freshwater on Earth is stored as groundwater. The amount of water held as groundwater is more than 100 times the amount collected in rivers and lakes (Shiklomanov and Rodda 2003). Most groundwater has accumulated over millions of years in vast aquifers located below the surface of the earth. These aquifers are replenished slowly by rainfall, with an average recharge rate that ranges from 0.1 to 3% per year (Covich 1993; La Gal et al. 2001). Assuming an average of 1% recharge rate, only 110×10^{12} m^3 of water per year are available for sustainable use worldwide. Currently, world groundwater aquifers provide approximately 23% of all water used throughout the world (USGS 2003). Irrigation for U.S. agriculture relies heavily upon groundwater, with 65% of irrigation water being pumped from aquifers (McCray 2001).

Population growth, increased irrigated agriculture, and other water uses are mining groundwater resources. Specifically, the uncontrolled rate of water withdrawal from aquifers is significantly faster than the natural rate of recharge, causing water tables to fall by more than 30 m in some U.S. regions (Brown 2002). The overdraft of global groundwater is estimated to be about 200×10^9 m^3 or nearly twice the average recharge rate according to the International Water Management Institute (2001).

The rapid depletion of groundwater poses a serious threat to water supplies in world agricultural regions especially for irrigated lands. Furthermore, when aquifers are mined, the surface soil area is prone to collapse, resulting in an aquifer that cannot be refilled (Youngquist 1997; Glennon 2002).

When atmospheric precipitation reaches the ground it divides into several sections which pursue the terrestrial part of the hydrological cycle along different paths. Many authors (Rockström 1999; Shiklamanov 2000; Ringersma et al. 2003), estimated that out of the total annual amount of precipitation of 110 km^3, about 40,000 km^3 is converted into surface runoff and aquifer recharge (blue water) and an estimated 70,000 km^3 is stored in the soil and later returns to the atmosphere through evaporation and plant transpiration (green water). Blue water is the freshwater that sustains aquatic ecosystem in rivers and lakes. It is used for drinking or domestic purposes, for industry and hydropower, and to irrigated agriculture, where around 70% of this water source is used. Indeed, irrigation uses not only blue water but also green water, whereas rain-fed agriculture uses only green water. According to Ringersma et al. (2003) the green water/blue water concept has proven to be useful in providing a more comprehensive vision of the issues related to water management, particularly in the agriculture sector.

Comprehensive assessment of water management in agriculture estimates that crop production takes up 13% (9,000 km^3 yr^{-1}) of the green water delivered to soil by precipitation, the remaining 87% being used by the non-domesticated vegetal world, including forests and range land. Blue-water withdrawals are about 9% of the total blue-water sources (43,800 km^3) with 70% of withdrawals going to irrigation (2,700 km^3). Total evapotranspiration by irrigated agriculture is about 2,200 km^3 (2% of rain) of which 650 km^3 are directly from rain or green water and the remainder from irrigation water or blue water. Competition among different sectors for scarce water resources and increasing public concern on water quality for human, animal, and industrial consumption as well as recreational activities, have focussed more attention on water management in agriculture. As water resources shrink and competition from other sectors grows, agriculture faces a dual challenge: to produce more food with less water and to prevent the deterioration of water quality through contamination with soil runoff, nutrients, and agrochemicals.

A global analysis of the future water for food requirements clearly indicates the urgent need for a concerted effort of simultaneous development of irrigated and rainfed farming systems. Current estimates of future expansion of storage reservoirs for irrigation are conservative (WCD 2000). In this context, Shiklomanov (2000) estimated an increase in irrigated land from 253 million ha in 1995–330 million ha by 2025, with an increase of approximately 500 km^3 yr^{-1} of consumptive water use. The FAO, in its latest study on agriculture until 2015/2030 (FAO 2002a), estimates a lower irrigated area in 2025 (271 million ha) based on an annual growth rate of 0.6% (1997–2025), with growth primarily in developing countries.

Falkenmark and Rockström (2004) reported a continued modest irrigation expansion of 0.6% per year from 2025 to 2050, the increased consumptive blue-water use for irrigation in 2050 would amount to 600 km^3 yr^{-1}. These authors consider all blue-water losses through seepage in conveyance channels and drainage from cropland as re-usable within or downstream of an irrigation system.

Improved management of existing irrigation systems will certainly result in improved water productivity, enabling the production of more food per unit water consumed through 2050. As was pointed out by Falkenmark and Rockström (2004) that an additional 5,600 km^3 yr^{-1} of consumptive green water may be required to properly feed the world population in 2050. According to the FAO (2002b) and Shiklomanov (2000) green-water productivity in irrigated agriculture averages 1,500–2,500 m^3 t^{-1}. With an estimated average green-water productivity of 1,700 m^3 t^{-1} in 2050, the total contribution from irrigation to food production in 2050 as a result of water productivity increases will amount to 200 km^3 yr^{-1}. This would result in a total contribution of 800 km^3 t^{-1} (600 km^3 yr^{-1} from increased water withdrawals, and 200 km^3 yr^{-1} from water productivity improvements) (Rockström and Barron 2007).

Therefore, this additional green water required for food production in 2050 can only come from expansion and production increase in rainfed farming systems.

A growing global population and the risk of crop failure caused by irregular climate are putting agricultural activities under pressure to meet increasing demand for food, feed, fibre, and fuel.

For agriculture to remain sustainable, it is essential to achieve the critical balance between improved productivity and environmental protection. Biodiversity, soil and water conservation, the welfare of rural communities, and the long-term success of human activities all depend on sustainable agriculture.

It is estimated that by the year 2030, the world population will have increased by some 1.7 billion people (UN 2004). Intensification enabled by new technologies has helped agriculture meet the world's demand for food and preserved natural habitats by slowing the expansion of land used for agriculture. The input of agricultural products must be used in conjunction with sustainable farming practices. Degradation of valuable natural resources today has direct implications for agricultural productivity in the future through soil erosion, water depletion, and the biodiversity loss.

Farmers play a significant role in managing ecosystems and protecting biodiversity. Inappropriate use of agrochemicals can pollute waterways, disrupt ecosystems and pose a risk to human health.

In this sense, the key objectives for meeting sustainable agriculture include:

- Improvement the water-use efficiency for agricultural crops.
- Minimize the risk of salinization and pollution of water bodies by agricultural inputs.
- Improvement the rainfall-water-harvesting techniques in rainfed farmlands.
- Protect and enhance the environment and natural resources.
- Protect the economic viability of farming operations.
- Provide sufficient financial reward to the farmer to enable continued production and contribute to the well-being of the sector.
- Produce sufficient high-quality and safe food.
- Build on available technology, knowledge, and skills in ways that suit local conditions and capacity.

If agriculture is to be sustainable, farmers must be able to make a net profit over the long term, and the resources on which farming depends must be used in a sustainable manner. Achieving long-term profit requires: (i) maintaining or improving resources; (ii) meeting legislative requirements with respect to the environment; and (iii) adjusting to international and local market demands.

Irrigation is a farming activity that has the potential to increase farmers' profit and to enhance the condition of natural resources, while having minimal impact on others. This project develops indicators of irrigation performance that can be used to measure and demonstrate the impacts of irrigation, and through their use, allow farmers to make informed management choices.

Sustainability goals for irrigated agriculture are shown in Table 6, followed by a list of indicators, which may be used to show progress towards these goals (MAF 1997).

Table 6 Sustainability goals and indicators

Type	Goals	Indicators
Overall	Maximise net profit over the long period	
Economic	Optimise farm productivity Maintain contribution to the wider economy	Annual net operating profit after tax (€ or $), quantity of crop/product produced per unit of water used (t m^{-3}), profit per unit of water used (€ or $ m^{-3}), quantity produced/hectare for each crop or product (t ha^{-1}), quality of produce (% of each crop or product at each grading level), annual energy used to operate an irrigation system (kWh), Energy used per volume of water pumped (kWh m^{-3}), and labour units per irrigated area (h ha^{-1})
Environmental	Hold and comply with resource consents Improve soil health Minimise adverse effects on water sources and receiving waters Minimise adverse effects on air Maintain or enhance biodiversity, habitats and landscape Pursue effective waste management Minimise use of non-renewable energy resources	Soil health (evaluated from soil water holding capacity, total organic nitrogen and carbon, pH and conditions of soil surface aggregates), daily volumes of irrigation water flowing onto farm for each crop (m3), daily % of water flowing onto the farm that is stored in the root zone (derived from soil-moisture measurements in and below the root zone), daily visual assessment of the amount of ponding or surface-water runoff, maximum water-extraction rate each season (m^3 day^{-1}), lysimeter-based measurement of nitrogen leaching below the root zone (effluent irrigation only), agrichemicals and fertilisers used per quantity of crop produced (kg ha^{-1} or L ha^{-1})
Social	Ensure acceptability of farming practices to the wider community Demonstrate good environmental management in the market place	Record of any abatement notices

4 Sustainable Agriculture and Climate Change

Global climate change may have serious impact on water resources and agriculture in the future. Therefore, numerous studies have been undertaken in recent decades to evaluate the ramifications of climate change for agriculture in various parts of the world (Fischer et al. 1996; Olesen and Bindi 2002; Aggarwal 2003; Jones and Thornton 2003). Based on a range of several current climate models, the mean annual global surface temperature is projected to increase by 1.4–5.8°C over the period of 1990–2100 (IPCC 2001a), with changes in the spatial and temporal patterns of precipitation (Southworth et al. 2000; Räisänen 2001). Arid and semiarid areas already suffering from limited availability of water under current conditions are likely to be most sensitive to climate change, while (sub-) wet areas may be less adversely affected as reported by Brumbelow and Georgakakos (2001) and Fuhrer (2003). Though different in socio-economic development, technological possibilities, and climatic regimes, the semi-arid regions that appear to have relatively ample water supplies for agriculture under the current climate are all most likely to be adversely affected due to greater water demand for irrigation projected under a warmer climate (IPCC 2001b; Rosenzweig et al. 2004).

Water-deficit stress can occur when precipitation does not adequately compensate for an increased evaporative demand due to a temperature rise. According to Haskett et al. (2000), this stress could cause declining yields or require more irrigation to maintain yields. This negative effect of increased temperature may be counteracted by effects of elevated CO_2 on crop tolerance to water stress (Lawlor and Mitchell 2000; Rosenzweig et al. 2004). Increased atmospheric CO_2 levels have major physiological effects on crop plants such as an accelerated photosynthetic rate. Depending on the inclusion and exclusion of CO_2-fertilization effect, as gain or loss is expected in crop yields (Haskett et al. 2000). As stated by Richter and Semenov (2005), scenarios simulated for the 2020s and 2050s showed that wheat yields in England are likely to increase more by the 2020s than in the following 30 years despite increasing CO_2 and temperature.

The history of agriculture reflects a series of adaptations to a wide range of factors from both within and outside agricultural systems. For instance, environmental conditions related to soil, water, terrain, and climate impose constraints and offer opportunities for agricultural production.

Also, technological developments lead to modifications in the structure and processes of farming operations.

In particular, weather and climate conditions have long been recognized as key determinants for success in the agro-food sector. Lobell and Asner (2003), analysing data on crop yields, temperature, precipitation, and solar radiation, have concluded that effects from climate have been underrated and are often mistakenly attributed to management practices. Variations in conditions such as length of growing season, timing of frosts, heat accumulation, precipitation, evaporation, and moisture availability all influence production and therefore economic returns to producers and farming businesses. With the onset of greenhouse-gas-induced climate change, growing conditions and climate-related risks and opportunities are expected to change, and may already be changing (Rosenzweig et al. 2000).

Most climate-change scenarios indicate that the gradual temperature rise, accompanied with reduced precipitation and enhanced CO_2 concentration, noted over the last several decades, will likely continue (McCarthy et al. 2001). However, these features tend to mask the projected changes in climate and weather conditions that producers cite as presenting the greatest risks, namely changes in the frequency, severity, and extent of extreme events (e.g. extended droughts and torrential rains) accompanied by intensified variability (Chiotti et al. 1997;

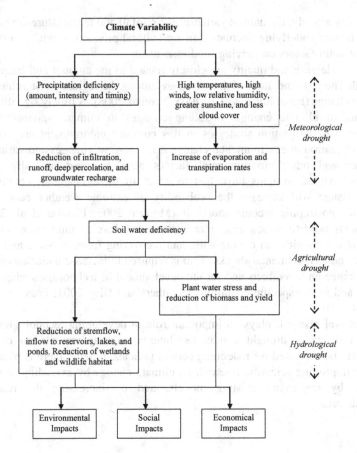

Fig. 7 Types of droughts and their impact

Reilly et al. 2001; Smit et al. 2000a). Figure 7 shows the impact of climate variability on the environment under a scenario of climate change. The capacity of a farming system to adapt to changing climate and weather conditions is based on its natural-resource endowment and associated conditions (e.g. economic, social, cultural, and political). The viability of these elements also constitutes the basis for sustainable agriculture, understood as agricultural production that: ensures adequacy of food production, does not harm the resource base, is economically viable, and enhances the quality of life (Smit and Smithers 1994).

An assessment of relevant strategies suggests that the two are closely aligned. Many climate and weather risk-management strategies fit squarely into sustainable agriculture practices and can, therefore, be promoted with several of the programmes and policies that target environmentally responsible production.

Climate change has evolved from a complex environmental issue to an even more complex developmental issue. Climate change is not a peripheral issue for development. This is especially true for the arid and semi-arid regions of the world. Today already, the natural variability in rainfall and temperature are among the main factors underlying fluctuation in agricultural production, which in turn is one of the main factors underlying food insecurity.

Water availability and quality are closely related to the amount and frequency of rainfall. The dry-land areas of the world are among the regions most vulnerable to climate change (Lioubimtseva et al. 2005; Thomas 2008). A timely signalling of the impact of climate change, including changes in climate variability and identification of adaptation strategies in this complex environment are crucial. Clearly, adaptation to environmental change is not new, as changes and variations in climate and other environmental factors have occurred naturally. Both human and natural systems have had to adapt to these changing conditions. Climate change will increase the probability of extreme weather conditions, leading to catastrophic income shortfalls (Monirul 2003; Planton et al. 2008). Governments need to review past interventions and develop innovative ways to assist rural communities in coping with, and recovering from, massive and large economic and environmental shocks. That is required to increase understanding of climate change and its effects and for the development of technologies adapted to location- and sector-specific conditions (Smithers and Blay 2001; Paavola 2008; Ford et al. 2010).

Agricultural research plays an important role in developing technologies that perform well under drought conditions. International agreements on climate change may be exploited for redefining certain policies. Finally, there is plenty of scope for improving scientific research on climate change by extending research networks, by improving existing models, and by increasing the research geographic area.

5 Irrigation and Sustainable Agriculture

The global water cycle, land management, and food security are intimately linked (Lal 2007; Hoff 2009; De Fraiture et al. 2010). The global food system has responded to the doubling of world population by more than doubling food production during the past five decades. Feeding a growing and wealthier population poses significant challenges for food security and environmental sustainability in the coming decades (Abdullah 2006). Producing more food requires more water; richer and more nutritious diets require even more water. Much of the additional food production must come from the intensification of land and water systems (FAO 2003; Khan et al. 2006). But this would exert unprecedented pressure on ecosystems which provide a range of benefits to mankind including food, fibre, timber, fuels, climate regulation, biodiversity conservation, and regulation of water flows and quality.

Sustainable irrigation means applying the correct amount of water at the appropriate time for optimal conditions of crop growth, minimizing overwatering, leaching, and runoff. Improving land and water management in agriculture and the livelihoods of the regional communities requires mitigating or preventing land degradation. Unsustainable land and water management can compromise the capacity of the ecosystems to provide livelihood and support services to mankind (Pimentel et al. 2004). According to Fujimori and Matsuoka (2007) the intensification of agricultural production has already doubled the amount of nitrogen sequestered globally and tripled the phosphorus use. This has led to eutrophication of lakes and coastal catchments, damaging fisheries, reducing recreational values and increasing the occurrence of toxic algae blooms (Hendry et al. 2006). Other negative impacts on the ecosystem services include loss of biodiversity, loss of pollinator species, and the flourishing of invasive, non-native species (Dudgeon 2000; Thrupp 2000). The expansion of agriculture and conversion of forests into cropland can alter biogeochemical cycles, including C sequestration capacity and hydrology (Ramankutty et al. 2002). Pesticide levels in surface water pose a health risk (Pingali et al. 1994); groundwater contamination can impair water quality and compromise community water supplies (Giraldez and Fox 1995). Other negative impacts attributed to unsustainable land and water use for agriculture include waterlogging and salinity-related losses in agricultural productivity (Murgai et al. 2001; Hussain et al. 2004), risks to public health and infrastructural damages from contaminated water supplies (Ragan et al. 2000), surface runoff of nutrients and agrochemicals causing euthophication and river health risks from excess river water withdrawals for irrigation (Reid and Brooks 2000; Saiko and Zonn 2000).

Degradation of land and water resources are commonly reported problems in large-scale irrigation systems. Widely reported data suggest that about 1/3 of global irrigated land has lower productivity due to poorly managed irrigation, causing waterlogging and salinity. Annually about 10 Mha is lost to salinization, of which about 1.5 Mha is irrigated lands. Global productivity loss due to land degradation over three decades has been estimated at 12% of total production from irrigated, rainfed, and rangeland or about 0.4% per annum cumulatively (World Bank 2003).

Therefore, unsustainable water management has implications for the sustainability of food production and of terrestrial and aquatic ecosystems as well as the services they provide to the humans (Tilman et al. 2002; Gisladottir and Stocking 2005). Often the potential adverse impact of irrigation is due to irrigation water per se, but rather to inadequate institutional and management responses such that many of these can be addressed with effective policies and programmes (Easter 1993). However, the links among sustainable land and water-use practices and ecosystem services on one hand and policies that can make these practices more sustainable on the other hand remain poorly understood.

In the context of the sustainability of irrigated agriculture, the following "sustainability targets" have been identified as being the key components of a sustainable agricultural system. A management practice such as applying fertiliser might help maintain soil productivity but can cause adverse effects on groundwater. Sustainable management includes the need to strike a balance between conflicting interests (Wichelns and Oster 2006; Calzadilla et al. 2010).

Table 7 Sustainability targets

Impact	Targets
Overall	Maximise net profit over the long term
Economic	Optimise farm productivity
	Maintain contribution to the wider economy
Environmental	Preserve and protect natural water resource
	Improve soil health
	Minimise adverse effects on water source and receiving waters
	Minimise adverse effects on air
	Maintain or enhance biodiversity, habitats, and landscape
	Pursue effective waste management
	Minimise use of non-renewable energy resources
Social	Ensure acceptability of farming practices to the wider community
	Demonstrate good environmental management on the market scale

It is recognised that maximising long-term net profit is the overall target for farmers, but that this target cannot continue to be met at the expense of the other economic, environmental, and social concerns identified (Table 7).

Irrigation has many far-reaching effects on the environment that may not be apparent at first, so it is important that from the beginning, the effects on the whole system should be addressed. It is therefore necessary to consider the relationship between these targets; considering any one target in isolation will lead to an overall system that does not meet the overriding principle of sustainable management.

Sustainable irrigation management builds on the concept that sustained improvements in the quality of human activities are possible only where the level and patterns of resource use are compatible with the natural environment and societal preferences for production process. The carrying-capacity-based irrigation management thus involves the closely integration of social expectations, ecological capabilities, and production, supporting the role of the resource by explicitly recognizing the spatio-temporal distribution of water resources (Khan and Hanjra 2008). Therefore, the operational framework for sustainable land- and water-management decisions thus have links between supportive capacity, assimilative capacity, optimal allocation of resources, and technological interventions, as both supportive and assimilative capacity can be enhanced through the new technological advances (Khan and Hanjra 2008). In addition, the carrying-capacity-based irrigation management includes mitigating negative water quantity and quality externalities that may involve options relating optimal irrigation volume, timing, and quality (Dinar and Zilberman 1991); irrigation and drainage reduction technologies (Dinar et al. 1992) and incentive policies (Wichelns 2002); investments in land and water resource knowledge (Dinar and Xepapadeas 1998) and farmer training (van Asten et al. 2004); joint management of surface waters and groundwater aquifers (Zeitouni and Dinar 1997); integrating environmental and water policies (Dinar and Howitt 1997); and cross-sectoral approaches such as input pricing policies, such as energy pricing especially for groundwater management (Scott and Shah 2004).

Appropriate irrigation management for economic as well as for environmental sustainability can be described as the best management policy (Boland et al. 2006). In particular, orchard irrigation involves many factors, including irrigation scheduling, nutrient management, salinity, and water-table control, vigour management using deficit-irrigation strategies and knowledge of the crop's critical phenological periods (Boland et al. 2001, 2002). Oster and Wichelns (2003) reported considerable background information on best management practices for irrigation that has evolved from many years of research and development in concrete areas of water management. Many monitoring programs have been conducted to measure the key indicators for sustainable irrigation practices and verify their application in the orchard (Boland et al. 1998).

5.1 Deficit Irrigation as a Sustainable Strategy for Optimising the Agricultural Use

The deficit irrigation is an irrigation practice by which the amount of supplementary water applied as irrigation is reduced to only a fraction of potential evapotranspiration from a well-watered reference crop (ET_C). According to English and Raja (1996) deficit irrigation is an optimising strategy under which crops are deliberately allowed to sustain some degree of water deficit and yield reduction. Its adoption implies knowledge of crop ET, crop responses to water deficits, including the identification of critical crop-growth periods, and the economic impact on yield-reduction strategies.

There are different ways for implementing an irrigation deficit strategy, differing mainly in how the water restriction is applied. Sustainable-deficit irrigation is based on a uniform application of a water restriction, depending on the crop-water demand. This strategy allows the crop to develop an adaptation to the stressful situation. Nevertheless, this approach does not consider the possible incidence of critical periods or the crop physiological status, in terms of crop-water availability. Under a sustainable-deficit-irrigation regime, the differential sensitivity of expansive growth and photosynthesis to water deficits leads to reduced biomass production under moderate water stress, due to a reduction in canopy size and in radiation interception (Fereres and Soriano 2007).

Regulated deficit irrigation strategy is based on applying different degrees of water stress in terms of crop phenological periods, dealing with this to minimize the effects on final yield or getting some benefits in reference to the final quality of the harvest product or crop development (Zhang et al. 2006; Bekele and Tilahun 2007; García-Tejero et al. 2010a, b, c; Hueso and Cuevas 2010). Such strategies seek to avoid as far as possible the application of water stress during the most determinant phenological periods for crop production, avoiding significant declines in yield and achieving significant increases in water productivity. A deficit-irrigation programme in non-critical periods is one of the cores of such strategies, allowing water stress within the crop tolera; nce range. It is vital to determine the water-deficit limit in each non-critical stage, to avoid potentially stunting crop development. It is not easy to determine the specific critical periods in each crop or the best strategy to follow. In many cases, it depends on crop, variety, and agro-climatic conditions, the experimentation being necessary as a basis for determining the range of water-stress tolerance on the different developmental stage.

Another deficit-irrigation strategy is based on the application of irrigation-restriction cycles while keeping the crop within a tolerance range (Enciso et al. 2003; Therios 2009). This strategy is known as low-frequency deficit irrigation, and requires continuous monitoring of the crop's water status. Under this strategy, the soil is left to become dry, since all the available water is lost to evapotranspiration. Consequently, the crop is irrigated to field capacity and again is left to dry. The main problem of this strategy lies in the establishment of a water-stress tolerance range for each crop. On the other hand, this strategy has the advantage of allowing a partial recovery of crop-water status. Muriel et al. (2009), studying a sweet orange orchard cv. Navelino, reported that low-frequency deficit irrigation in comparison to sustainable-deficit irrigation provided better results regarding yield and fruit quality, with similar water inputs (60% ET_C.). Although the accumulated water-stress levels over the irrigation period were similar in both treatments, the partial recovery of crop-water status in the low-frequency deficit irrigation gave yield values 20% higher than those in the sustainable-deficit-irrigation treatment. One of the main effects was detected in the fruit-growth slope. Meanwhile, it was relatively constant on sustainable-deficit irrigation and some 30% lower than that recorded in fruits of fully irrigated trees (100% ET_c.). However, during the irrigation cycles, there was a recovery of fruit growth slope, being even higher than recorded in full irrigated trees, which allowed a nearly complete recovery of fruit size.

Finally, a partial root-zone drying system is a deficit-irrigation strategy based on the irrigation of 50% of the root-system, while the other half is dry, in any given period. This strategy alternates drying and wetting cycles to roots, and allows a plant growth with a reduced transpiration, without chemical signals of water stress (Kang and Zhang 2004; Santos et al. 2003). This technique significantly reduces field-crop and fruit-tree water use, increases canopy vigour and maintains crop yield. This technique takes into account the physiological processes promoted by soil drying. That is, a bio-chemical response occurs based on abscisic acid synthesis by the roots in the drying soil, and this hormone is transported through the transpiration process to the stomata (Zhang et al. 1987; Zhang and Davies 1989, 1990; Liang et al. 1997). This signal leads to partial stomata closure, reducing the water loss (Davies and Zhang 1991).

In conclusion, a correct strategy of deficit irrigation is the basis to maximize water productivity. Although a relative yield reduction can be caused in areas where water availability is the main limiting factor for the crop, water saving throughout the application of deficit irrigation can be economically more profitable

Table 8 Differences and similarities between regulated deficit irrigation and partial root-zone drying

Regulated deficit irrigation	Partial root-zone drying
Confirmed both in fruit crops and wine grapes	Confirmed for wine grapes, potential for fruit crops
Irrigation saving assured	
Fruit crops maintain final size and yield	Potential for fruit crops and research continues
Wine grapes produce smaller berries with reduced yield	Wine grapes maintain berry size and yield
The timing is critical	The timing is flexible
Improvement of quality in wine grapes and fruit crops	Potential for improved quality via reduced vigour in wine grapes
Reduce vegetative vigour conducive to improved cropping	
Deficit irrigation where only uppermost profile is re-wetted	Deficit irrigation where deeper wet/dry zones are spatially separated

than maximizing the yield, thereby guaranteeing the viability of the agro-eco-systems and the economic profitability for the farmers (Geerts and Raes 2009). Moreover, the slightly impact on yield can be mitigated by improving some fruit-quality properties (total soluble solids, titrable acidity, fruit or grain size among others (García-Tejero et al. 2010a, b; Zhang et al. 2008; Spreer et al. 2009; Cui et al. 2008; Di Paolo and Rinaldi 2008; Farré and Faci 2006).

Principal features of regulated deficit irrigation and partial root-zone drying system that highlight some key distinctions between them are show in Table 8.

5.1.1 Irrigation Scheduling

Irrigation requires a relatively high investment in equipment, fuel, maintenance, and labour, but offers a significant potential for increasing net farm income. Frequency and timing of water applications have a major impact on crop yields and operating costs. To schedule irrigation for most efficient use of water and to optimise crop yield, it is desirable to frequently determine the soil–water conditions throughout the root zone of the crop being grown (Pereira 1999; Imtiyaz et al. 2000; Bergez et al. 2002; Li et al. 2005). A number of methods for doing this have been developed and used with varying degrees of success. In comparison to investment in irrigation equipment, these scheduling methods are relatively inexpensive. When properly used and coupled with grower experience, a scheduling method can improve the irrigator's chances of success.

Therefore, the objective of irrigation scheduling is to define *how*, *when* and *how much* to irrigate a crop, based on the water needs, providing the optimal amount for the development of the plant. The need for a good irrigation planning is especially important under situations where the water scarcity is a limiting factor or when the water demand is over the potentially available reserves. In this sense, is essential to quantify the crop's water requirements in order to avoid unnecessary contributions that promote water losses from runoff or deep percolation. An appropriate irrigation scheduling helps maximize net return, minimize irrigation costs, maximize yield, optimally distribute limited water, and minimize groundwater pollution (Huygen et al. 1995; Pereira 1999; Imtiyaz et al. 2000).

The crop, through the photosynthesis process produces biomass, using solar energy, atmospheric CO_2, water, and mineral salts from the soil solution. The CO_2 fixation and exchange require the opening of stomata, regulating the input and the output of water vapour in what is known as the crop transpiration, which is linked to atmospheric conditions (San Jose et al. 2007; Qin et al. 2010). This process of water extraction from soil to atmosphere is owed to a simple diffusion mechanism, promoted by the potential difference between soil, plant tissues and the atmosphere. However, the daily transmission of water to the atmosphere in this process may exceed the plants' water content several-fold. Therefore, many crops retain less than 1% of the water they take up from the soil (Hillel 2008). The rest of water transpired is released into the atmosphere through the stomata.

On the other hand, there is a large amount of water lost directly by evaporation from the soil surface. All of the water lost by transpiration and evaporation is referred to as the crop evapotranspiration (ET_c), defined as the total vapour water loss of a green cover, through evaporation and transpiration, in a time interval, which depends mainly on the type of crop, its phenological stage, the degree of coverage and the climatic conditions (Allen et al. 1998). ET_c determines the actual water needs of a crop, the latter being understood as the volume of water required for a crop from the start of season until harvest, so that there is no a limitation on growth or development, nor repercussions on crop yield. The calculation of ETc values is the basis of any method for scheduling irrigation, since this establishes the water amount needed by the crop. Therefore, good irrigation scheduling means applying the right amount of water at the right time, making sure water is available when the crop needs it, maximizing the water efficiency while minimizing runoff and percolation losses.

Crop water use can be estimated by several methods: weather data, soil-moisture sensors or monitoring plant stress. Whereas the first of these is based on the calculation of ET_c, soil-moisture sensors can indicate the soil–water content and from a water level close to field capacity, monitoring the spatial and temporal variability of soil moisture. On other hand, the use of plant sensors are based on monitoring the crop-water status, and its use reveals exactly at all times the physiological status of the plant and the possible existence of water stress.

Weather Data

There are many methods for estimating ET_c, which can be divided into methods based on the calculation of a reference evapotranspiration (ET_0) and the application of a crop coefficient (K_c); and the direct method of weighing or of drainage lysimeters.

The concept of the reference evapotranspiration was introduced to study the evaporative demand of the atmosphere, independently of the crop type, crop development, and management practices. The only parameters that affect ET_0 are the climatic variables, and therefore it can be calculated from weather data. Several methods allow the calculation of ET_0: Blanney and Criddle (1950); Blanney-Criddle FAO (Allen and Pruitt 1986), Hargreaves method (Allen et al. 1998), although the FAO Penman–Monteith method (Allen et al. 1998) is the most used for determining the reference ET_0. This method requires some meteorological data such as radiation, temperature, air humidity, and wind speed, using the following equation:

$$ET_0 = \frac{0.408\Delta(R_n - G) + \gamma \frac{900}{T+273} u_2 (e_s - e_a)}{\Delta + \gamma(1 + 0.34 u_2)}, \tag{5}$$

where ET_0 is the reference evapotranspiration (mm day^{-1}); R_n is the net radiation at the crop surface (MJ m^{-2} day^{-1}); G is the soil heat flux density (MJ m^{-2} day^{-1}); T is the mean daily temperature at 2 m height (°C); u_2 is the wind speed at 2 m height (m s^{-1}); e_s is the saturation vapour pressure (KPa); e_a is the actual vapour pressure (KPa); Δ is the slope vapour pressure curve (KPa°C^{-1}); and γ is the psychrometric constant (KPa°C^{-1}).

Alternatively, the reference ET_0 can be estimated by the Pan evaporation Method (Allen et al. 1998). This procedure relates the pan evaporation to the ET_0 by an empirically derived pan coefficient, according to the equation:

$$ET_0 = k_p E_{pan}, \tag{6}$$

where ET_0 is the reference evapotranspiration (mm day^{-1}); k_p is the pan coefficient, and it depends on the type of pan, size and location, among other factors; and E_{pan} is the pan evaporation (mm day^{-1}).

Secondly, ET_c is related to the reference ET_0 through a K_c, which depends on the type of crop, its physiological state, the type of management, and the development of plant cover. Thus, whereas the ET_0 depends entirely on the weather conditions reported, the crop coefficient fits the reference evapotranspiration to the specific crop conditions. Therefore, ET_c is calculated by the equation proposed by Doorenbos and Pruitt (1977):

$$ET_c = ET_0 \cdot k_c, \tag{7}$$

where ET_c is the crop evapotranspiration (mm day^{-1}); ET_0 is the reference evapotranspiration (mm day^{-1}); and K_c is the crop coefficient.

While the reference ET_0 calculation has been widely studied, with several methods for its calculation (Allen et al. 1998), there is a great uncertainty for K_c values (Fereres 1996). Doorenbos and Pruitt (1977) proposed generic K_c values that vary depending on the crop type, management or stage development. However, these values are highly dependent on the crop location and the climatic conditions. Therefore, heat advection can augment the evaporative demand of crops (Brakke et al. 1978). In addition, stomatal opening depends on the values of vapour-pressure deficit, causing variations in the rates of transpiration for a specific crop (Fereres 1984).

On the other hand, ET_c can be directly measured by the lysimeter method. A lysimeter is an instrument that measures water movement in soils (Howell et al. 1991), there are two types: weighing and drainage. With drainage lysimeters, changes in soil–water content are estimated indirectly. With weighing lysimeters, changes in soil water within a constructed container are measured (Seyfried et al. 2001). Through a balance of input–output, the ET_c can be estimated.

An appropriate irrigation schedule should consider not only the amount of water demanded by the crop, but also timing and the irrigation intervals for general conditions of the soil–plant-atmosphere system. Thus, ET_c offers information only concerning the crop water demand, but does not take into account the soil–water content or the crop-water status.

Soil–Water Content

Soil acts as a dynamic storage system, retaining water and returning it to the atmosphere (i) from the soil surface by evaporation and (ii) from the plant surface by transpiration. The soil–water content directly affects plant growth as it determines the plant-water status. Irrigation scheduling can be established using several different strategies based on soil–water (Huygen et al. 1995). There are two ways to assess soil–water availability for plant growth: by measuring the soil–water content and the soil–water potential (Itier et al. 2010).

For irrigation purposes, soil–water content is expressed as a fraction of the available water, which is given by the ratio of available water content to available water capacity, which is defined by water contents at field capacity and at the wilting point (Fig. 8).

Soil moisture can be determined through the thermo-gravimetric method, by the difference in wet and dry weight of a sample, calculating the relationship between the weight of stored water and the total weight of dry sample (θ_g, kg kg^{-1}). This parameter can be expressed in relation to the volume of water retained per unit of soil volume, using the proceeds of θ_g by bulk density (ρb, Mg m^{-3}). Although this method allows for very accurate data, it is highly time consuming for the analysis of soil–water balance and irrigation management.

Among the methods currently used for continuous monitoring the soil–water content are the dielectric methods based on the measurement of some dielectric soil properties, which depend heavily on soil moisture (Verhoef et al. 2006; Frangi et al. 2009; Sagnard et al. 2009). These tools are not new for the calculation of crop-water requirements, although its use in irrigation scheduling is quite limited. The application of these techniques is to reduce the excess of water inputs that occur with the use of empirical methods based solely on estimates the crop evapotranspiration.

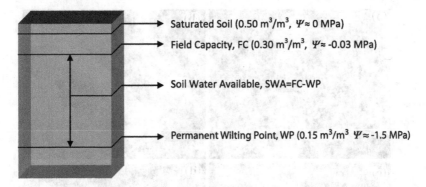

Fig. 8 Different soil–water levels for irrigation scheduling

Among these, time-domain reflectometry (TDR) sensors measure the soil dielectric constant (ε_b) through the transmission time of a high-frequency electromagnetic pulse, which propagates along a conductor embedded in the soil (Ledieu et al. 1986; Noborio 2001; Ferré and Topp 2002). Thus, the propagation velocity (v) is a function of εb, and is proportional to the square of time transit (t) back and forth along the conducting medium. These sensors measure within a moisture range of 0.05–0.5 m^{-3}, with a reading error about 2%, although in certain soil types, local calibrations are recommended (Evett 1998).

On the other hand, frequency domain reflectometry (FDR) or capacitance sensors estimate the soil–water content, taking into account the response to changes in the dielectric constant of soil, using a domain reflectometry technique known as capacitance (Brandelik and Hübner 1996; Bilskie 1997; Laboski et al. 2001). Capacitance sensors measure the dielectric permittivity of the medium through the charging of a capacitor which is in contact with the soil by an access tube (Fig. 9). When an electric field is applied, the soil contact with the electrodes acts as the dielectric of a capacitor. In electromagnetic terms, a soil is characterized by 4 components: air, solid phase, not available, and available water (Hallikainen et al. 1985).

TDR and FDR probes offer great advantages over other systems, such as the possibility of obtaining a large number of measures, continuously, and without disturbing the soil properties. Both devices need to be calibrated at soil local conditions especially in soils with high salinity, high organic matter content or clayey soils (Sentek 1999; 2001).

Soil–water potential measurements are related to the force with which soil water is retained. These measurements give information about the extraction force that the plant needs to take up the water from the medium. The soil–water potential is the sum of several component potentials:

$$\Psi = \Psi m + \Psi o + \Psi p + \Psi g, \tag{8}$$

Fig. 9 Different devices for measurements soil and plant water status: Multi sensor capacitance probes (FDR) (**a**), dendrometer for fruit diameter (**b**), plant water potential (**c**), porometer for stomatal conductance (**d**), device for leaf photosynthetic activity (**e**), and sap flow and dendrometer for trunk fluctuations (**f**)

where Ψ is the potential energy per unit mass, volume, or weight of the water, and the subscripts, m, o, p, and g are for matric, osmotic, pressure, and gravitational potential components.

The soil–water potential is used as a tool for estimating the irrigation scheduling for many crops (Michelakis et al. 1996; Novák et al. 2005; Wang et al. 2007a). These types of sensors have a porous material that comes in contact with the medium through which water can freely circulate. Thus, when soil dries, it produces suction from inside the porous medium to the ground, and the opposite occurs when soil is rewetted.

A tensiometer measures the soil–water tension, which is related to the soil–water content and provide information related to irrigation needs (Krüger et al. 1999; Wang et al. 2007b; Merot et al. 2008). The basic structure of this system is a closed tube with a ceramic tip attached to the end, connected to a manometer. The tube is filled with water and sealed. As the soil dries, water is drawn from the tube, the manometer registering the suction force. When the ceramic tip comes into moisture equilibrium with the soil, the manometer registers the soil tension. These devices measure only the soil-matric potential down to −0.08 MPa. On the other hand, electric tensiometers are able to measure down to −0.2 MPa, and they have the advantage that the measurement can be logged (Malano et al. 1996).

Other types of sensors for monitoring soil–water content are the gypsum block (Fowler and Lopushinsky 1989; Stenitzer 1993). They consist of an electro-chemical cell formed by a pair of electrodes embedded in a porous capsule, and a saturated solution as electrolyte. They are very sensitive to temperature changes and variations in the soil electrical conductivity. Its measurement range is higher than that of the tensiometer (0.03–0.2 MPa), although it has a low resolution, especially when soil is nearly to saturation.

Plant–Water Status

The study of plant–water status is based on monitoring the behaviour of the tissues, water both in the variation of the moisture content as the force with which it is retained within the plant tissues. Such techniques have the advantage of providing information concerning the plant–water status as well for the possible existence of water stress (Patakas et al. 2005; Baeza et al. 2007). This information is useful for scheduling irrigation, but has the great disadvantage that it offers no information on the water required by the crop.

Many techniques are currently known, based mainly on the study of crop-water status, which can be classified among those which are based on discrete mea-surements (i.e. leaf- or stem-water potential and stomatal conductance), which are laborious but provide highly reliable data, and other based on plant physiological sensors (i.e. trunk-diameter fluctuations and sap-flow sensors), with the advantage of providing continuous measurements (Fig. 9).

It is possible to determine the plant–water status through the measurement of water potential (Ψ) (Schaffer and Whyley 2002; Baeza et al. 2007; Sato et al. 2007). Under optimal irrigation conditions, plants tend to maintain Ψ values close to zero, in order to maintain tissue turgidity. A decrease in Ψ is defined as an increase of water-retention force, or, the equivalent, a water-stress situation, whereupon the plant tends to retain water more strongly in its tissues (Salisbury and Ross 1985). This measure also provides information concerning the water movement through the soil–plant-atmosphere system. Consequently, water tends to move from areas with higher potential to others with lower potential. In fact, the main force involved in the evapotranspiration process is the difference in water potential between the soil, the plant, and the air surrounding the leaves.

These measurements can be made in any part of the plant, mainly in leaves (Ψ_{leaf}) or stem (Ψ_{stem}). Leaf–water potential have been widely used to determine the plant–water status and for irrigation scheduling. However, for this purpose, the stem water potential is being more commonly used, especially since Shackel et al. (2000) stated that these measurements were less influenced by the inherent variability of measurements. The Ψ_{leaf} and Ψ_{stem} measurements are made through pressure chambers (Scholander et al. 1964), following the methodology proposed by Turner (1988).

According to De Swaef et al. (2009), Ψ_{stem} directly reflects the plant's water status, bearing strong relationships with other physiological parameters such as sap-flow or radial-stem growth. In addition, many authors (Goldhamer et al. 1999; Naor and Cohen 2003; Nortes et al. 2005) found similar results for Ψ_{stem}, reporting the advantages of estimating this physiological parameter (Shackel et al. 1997; Naor 2000).

Stomatal conductance (g_S) can be defined as the inverse of the resistance offered by the stomata to the output of $H_2O_{(v)}$ on the leaf surface (mol m^{-2} s^{-1}). Stomata play a key role in plant physiology by controlling CO_2 fixation and water loss in the transpiration process. Stomatal conductance is highly dependent on environmental conditions such as radiation, air humidity, temperature, vapour-pressure deficit and soil–water content (Del-Pozo et al. 2005; Matsumoto et al. 2005). This regulatory capacity is due to two guard cells that open or close the stomatal pore, depending on these conditions. Thus, when weather conditions are strongly adverse, with high rates of evapotranspiration, or the soil–water content is low, the plant responds by closing the stomata or decreasing the degree of openness (Taiz and Zeiger 1998; Bray 1997). Sometimes it happens that, even when the plant has good hydration, if the vapour-pressure deficit is very high, the plant is unable to absorb water from the soil at the speed needed. In response, there is a prompt stomatal closure, as on the hottest summer days. The plant then must find a balance between its transpiration level, atmospheric demand and the amount of CO_2 necessary to develop the dark phase of photosynthesis (Bacon 2004). This parameter is measured through a porometer, which varies depending on the measurement method.

Comparatively, stomatal-conductance behaviour is generally not sensitive to changes up to a certain threshold of Ψ_{stem}. Moreover, this parameter is influenced by several factors, such as leaf morphology, irradiance, air temperature, relative humidity, or hormone synthesis (Smith and Hollinger 1991). Ortuño et al. (2004), have observed that fluctuations in g_S are higher than in Ψ_{stem}, in well-irrigated lemon trees, although the significant differences between two studied irrigation treatments were similar in both parameters. They observed that Ψ_{stem} was more sensitive to water stress than g_S, showing significant differences between two irrigation treatments one week before for Ψ_{stem} compared with g_S.

Water potential or stomatal conductance provides only punctual information on the crop–water status, this requiring a large number of measurements to compile enough information for proper decision making. For this reason, there is a set of

Fig. 10 Trunk-fluctuation
cycles

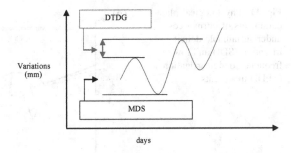

tools that provides continuous measurements in real time, guiding appropriate irrigation scheduling.

Dendrometry is a technique that allows continuous monitoring of stem- or fruit-diameter fluctuations, which are related to the crop–water status (Higgs and Jones 1984; Fernández and Cuevas 2010; Ortuño et al. 2010). Trunk and fruit diameters fluctuate diurnally in response to atmospheric conditions and changes in crop-water content (Fig. 9). Diurnal dynamics of diameter changes, especially of fruits, have been widely used as a sensitive indicator of irrigation need. The daily fluctuation cycle provides three different indices: maximum daily trunk diameter (MXTD); minimum daily trunk diameter (MNTD) and the maximum daily trunk shrinkage (MDS), calculated as the difference between MXTD and MNTD (Fig. 10). On other hand, daily trunk-diameter growth (DTDG) rate is calculated between the differences of MXTD on two consecutive days (Ortuño et al. 2004).

Many studies have suggested this type of sensor as a good tool for monitoring the crop-water status. Fereres and Goldhamer (2003) found that MDS was a better water-stress indicator than Ψ_{stem}, in almond trees. Moriana and Fereres (2002) showed that the differences between two consecutive maximum daily trunk diameter, was a good indicator for scheduling irrigation in young olive trees. In addition, Muriel et al. (2009) and García-Tejero et al. (2009, 2010c) demonstrated that MDS is a good water-stress indicator in mature citrus trees, indicating that these types of sensors offer reliable information, in real time, on physiological crop status, optimising a deficit-irrigation strategy based on crop-water status. Similar Ortuño et al. (2004), suggested MDS as a highly sensible indicator for young lemon trees than other water-stress parameters used with this technique such as MNTD or MXTD as well as other physiological parameters ($\Psi_{pre\text{-}dawn}$, Ψ_{midday}, and g_S). However, fluctuations in MDS are not related only to the crop-water status, but also respond sensitively to changes of several climatic parameters as daily mean vapour-pressure deficit, vapour-pressure deficit at midday, daily mean temperature, temperature at midday; ET_O or radiation. In these sense, Moreno et al. (2006) reported high correlations between MDS and some variables such as temperature and vapour-pressure deficit (both at midday). Fereres and Goldhamer (2003), studying young almond trees, showed MDS to be more sensitive to changes in mean daily vapour-pressure deficit, although Vélez (2004) indicated that MDS was more directly related to changes in radiation and ET_O than vapour-pressure deficit and temperature in clementina trees.

Fig. 11 Physiological plant
parameters of citrus trees
under sustainable-deficit
irrigation (SDI) and low-
frequency deficit irrigation
(LFDI) treatments

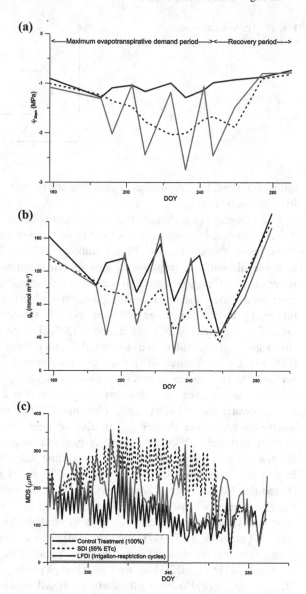

In this context, García-Tejero et al. (2010c) studied the temporary evolution of
different physiological plant parameters of citrus trees subjected to two deficit
treatments: sustainable-deficit irrigation (SDI) and low-frequency deficit irrigation
(LFDI) with similar amounts of water but with different timing. According to the
results, Ψ_{stem} remained relatively constant in the fully irrigated trees (100% ET_C), in
contrast to the remaining treatments that were subjected to water stress (Fig. 11).
The low-frequency deficit irrigation treatment offered a variable behaviour
according to plant–water status throughout the monitoring period. This is a

Table 9 Relationships between Ψ_{stem} and fruit parameters

	Ψ_{stem}	Yield	Fruit weight	TSS	TA	ED
Yield	−0.227*	1				
Fresh weight	−0.560**	0.166	1			
TSS	0.662**	−0.186	−0.637**	1		
TA	0.754**	−0.106	−0.595**	0.723**	1	
ED	−0.540**	0.129	0.960**	−0.614**	−0.565**	1

TSS Total solid soluble, *TA* Titrable acidity, *ED* Equatorial diameter
* $p < 0.05$
** $p < 0.01$

decreasing tendency of Ψ_{stem} during the restriction of irrigation. A similar trend was observed for the g_S, being more highly variable than Ψ_{stem} for the monitoring period.

Finally, the MDS in the trees studied depended on the amount of irrigation water applied in each treatment, values being relatively homogeneous in control trees (Fig. 11). However, the MDS in sustainable-deficit irrigation showed an increasing trend, while in low-frequency deficit irrigation treatment was highly variable, and similar to those found for Ψ_{stem} and g_S values.

In order to use these techniques for predicting the impact on fruit yield and quality it was studied the relationships between Ψ_{stem} and fruit parameters, being especially significant for fruit quality (Table 9).

Sap-flow measurements are a widely used method for estimate the crop-water use and storage in woody species (Wilson et al. 2001; Nicolas et al. 2005; Rousseaux et al. 2009). The most commonly used measurement methods are based on detecting convective heat transfer (heat carried with sap stream) inside tree trunks. There are three types of measurements, depending on the heating and signal-detection method. Heat balance and dissipation methods use the temperature difference of two or more sensors to calculate the flow rate. On the other hand, heat-pulse methods use the transfer rate of a heat pulse between two or more sensors as a signal for the sap-flow rate (Sevanto et al. 2009). Detailed reviews of these methods have been described by many authors (Campbell 1991; Cohen 1993; Čermák 1995; Granier 1985; González et al. 2008; Smith and Allen 1997; Green et al. 2003). However, among these methods, heat pulse methods offer accurate measurements of sap flow, with a minimal disruption to the sap stream. These measurements offer a good time resolution of sap flow, and enable automatic data collection and storage (Green et al. 2003) (Fig. 9). The compensation method uses two temperature probes placed on either side of a line heater that is inserted radially into the tree trunk. Following the application of a brief 1- to 2 s heat pulse, the time delay for an equal temperature rise at both sensors is used to calculate a heat-pulse velocity. On the other hand, a correction factor is used to correct heat pulse measurements for any probe-induced effects of wounding and to calculate rates of sap flow (Green et al. 2003). The measuring system (Green 1998) is made up of a set of probes and associated electronic components, connected to a data logger. Each probe set consists of a linear heater and two temperature sensors that are installed radially into the tree stem. The probe location depends on the method used (T-max method or compensation method; details of heat-pulse methods have been described by Green et al. (2003, 2009). Nevertheless, these methods are not able to measure low or reverse flows. Low flows are of particular interest at night because they can provide evidence of recharge of the plant's water status, and may also provide an indicator of plant-water stress (Fuentes et al. 2009). Several works have indicated that sap-flow measurements are good indicators for scheduling irrigation (Jones 2004; Nadezhdina 1997; Nadezhdina 1999; Fernández et al. 2008a, b) and evaluating the physiological effects of an application of water-deficit-irrigation strategy in several crops: Fernández et al. (2001) and Moreno et al. 1996b in olives; Romero et al. (2009) in almond; Green et al. (1998) in apple; and Alarcón et al. (2000) in apricot trees among others, determining the close relationship between sap-flow rates and water use.

In short, all these techniques need to be considered as potential solutions to improve agricultural water use. Each technique provides relevant information on some of elements of the soil–plant-atmosphere system. Therefore, an integrated management of agriculture water use requires the implementation of all needed tools to gather as much information of what is happening in the soil–plant-atmosphere system.

5.1.2 Impact of Deficit Irrigation on Water Productivity

Water is undoubtedly the most limiting factor in agriculture, especially in those arid and semiarid areas of the world where rains are below of ET_O rates. In such situations, crop development requires the contribution of water via irrigation to offset this imbalance. The problem lies in the fact that, in many cases, the available water reserves is below to the potential demand in agriculture. When this occurs, a search for different management strategies is needed to maintain the viability of agro-ecosystems, without compromising the profitability of farmer.

Water-productivity concept is based on the measurement of harvest product per unit of applied water. Between the techniques that aim to increase the water productivity is the use of management tools that enable more sustainable water use, such as those discussed in the previous sections. Nevertheless, in many cases the available water-use resources are not sufficient to bridge the demand, which seeks to create other strategies to achieve productive results within a tolerance range for the farmer. There is an increasing challenge for scientists to develop innovative soil–water-nutrient-crop-management practices that encourage sustainable agricultural systems (Anapalli et al. 2008), thereby maximizing water savings and improving productivity (Spreer et al. 2007; Gijón et al. 2009). Deficit irrigation has been widely investigated as a valuable and sustainable production strategy, especially in dry regions. In this context, many studies have reported the adoption the deficit irrigation in order to improve the water productivity in different crops (Ali et al. 2007; Geerts and Raes 2009; Singh et al. 2010; García-Tejero et al. 2010a, b, c; Egea et al. 2010). These practices save water by reducing crop evapotranspiration and increasing irrigation-water productivity (Fan et al. 2005). Such water-management methods exert different effects on the crop, altering the potential development of the plant, by depressing photosynthetic rates, reducing the sources of carbon (Hsiao 1973), and exerting a negative impact on the crop development and production (González and Castel 2000). Nevertheless, not all deficit irrigation strategies have these negative consequences. Water productivity is affected mainly by the yield response to deficit irrigation. It has been reported that it is not only biomass production that is linearly related to transpiration, but the yield of many crops is also linearly related to ET (Fereres and Soriano 2007).

Many authors have found that the response to water stress depends mainly on the crop phenology, and the different effects observed are closely related to the timing, duration, crop physiological status, irrigation-water quality, plant genotype, and the degree of stress endured by the crop (Doorenbos and Kassam 1979; Ginestar and Castel 1996; García-Tejero et al. 2008). The negative impact could be mitigated by improving fruit quality, as has been shown by several authors in the citrus crop (**Sánchez-Blanco et al. 1989, González and Castel 2000; Verreynne et al. 2001; García-Tejero et al. 2010a, b, c, pears (Mitchell et al. 1989), almond (Goldhamer and Viveros 2000, Romero et al. 2009), apple (Ebel et al. 2001), apricot (Ruiz et al. 2000), wine grapes (McCarthy et al. 2002) and olive (Moriana et al. 2003).

The final effects of deficit irrigation and thus the irrigation water productivity will depend mainly on the type of crop and the followed irrigation strategy.

Deciduous

Traditionally, the deficit-irrigation applied in these crops has been associated with the improvement of some aspects, gaining the benefits of either a decrease in vegetative growth or an improvement in some fruit-quality properties (Dos Santos et al. 2007; Acevedo et al. 2010; Egea et al. 2010). Its application has been developed primarily in those stages where the water stresses seemed not to offer problems in terms of yield or vegetative growth (Naor 2006). In this line, there have been numerous studies on the effects of water stress on crops, emphasizing the different effects depending on the degree on water stress applied and the development stage, although the results are highly variable.

In apple (*Mallus domestica* L.) many studies have concluded that the effects of deficit irrigation on yield and fruit size did not depend of the phenological stage when water stress is applied (Mpelasoka et al. 2001; Ebel et al. 2001). Leib et al. (2006) determined that a partial root-zone drying strategy with water saving of 25% did not affect yield or fruit size, compared with a control irrigation treatment (100% ETc).

However, other results show significant differences depending on the phenological stage in which a water deficit is applied Failla et al. (1992) showed that water stress during the reproductive cell-division stage promoted smaller fruits, and these effects persisted until the harvest. Similar effects were found when the water stress was applied during the post-reproductive cell-division stage (Mpelasoka et al. 2000, 2001). Regarding to the effects of water deficit in fruit-quality parameters in apple, several results show that deficit irrigation promotes increases ethylene synthesis, total soluble solids, and the proportion of mature fruits (Mpelasoka et al. 2000, 2001). Finally, Mpelasoka et al. (2001) and Mpelasoka and Behboudian (2002) reported relations between water stress and flesh firmness, although this finding could be an effect of fruit-size decrease as a result of deficit irrigation, since, according to Ebel et al. (1993) fruit firmness is directly related to lower fruit weight.

Similar results have been found in pear (*Pyrus communis* L.). Kang et al. (2002), testing a partial root-zone-drying strategy, with water savings approaching 25% compared with trees irrigated at 100% ET_C, detected no significant effects on yield or fruit quality. On the other hand, water stress imposed during the reproductive cell-division stage promoted a decline in vegetative growth, without limiting fruit size (Mitchell et al. 1986, 1989). Marsal et al. (2000, 2002) showed a direct relationship between the fruit size and the Ψ_{stem}. They pointed out that water stress during this phenological stage resulted in a fruit-size reduction, and this effect persisted until harvest, results in line with other studies (Naor et al. 2000, 2006). Other authors reported a closely relationship between the degree of water stress and fruit number, when a deficit-irrigation strategy was applied during the cell-division stage and/or postharvest stage (Mitchell et al. 1989). However, when the deficit irrigation was applied during the fruit-growth period, fruit size was significantly reduced (Marsal et al. 2000). By contrast, other authors have reported no significant effects on fruit size, although significant effects were found in fruit weight (Mpelasoka et al. 2001; Behboudian et al. 1998).

Finally, deficit irrigation did not clearly affect fruit quality. Unlike the data reported for apple, water stress did not affect fruit firmness (Ramos et al. 1994) nor total soluble solids (Behboudian et al. 1994), although these appear to be related to water stress with an increase in the proportion of mature fruits (Caspari et al. 1996).

Deficit irrigation effects have been widely studied in stone fruits (peach, nectarine, apricot or almond among others), and many studies have shown that a moderate water-stress level during the cell-division stage significantly diminished fruit size, although the loss was recovered by irrigation during the fruit-growth period (Goldhamer et al. 2002; Girona et al. 2002, 2004; Ruiz et al. 2000; Torrecillas et al. 2000). On other hand, moderate water stress applied during the final fruit growth stage significantly decreases fruit size and fruit weight in nectarine (Naor et al. 2001), Japanese plum (Naor et al. 2004), peach (Girona et al. 2002), and apricot (Torrecillas et al. 2000). Regarding fruit-quality parameters, Gelly et al. (2003) indicated that deficit irrigation improved fruit firmness and ethylene concentrations in peach fruits. Li et al. (1989) found higher soluble solids contents for peaches under water stress in different phenological periods. Torrecillas et al. (2000) analysed different quality parameters in apricots, but did not observe an effect of deficit irrigation on fruit quality.

Almond trees (*Prunus dulcis* Mill. D.A. Webb) represent one of the most important crops in many arid and semi-arid areas, having been traditionally associated with unproductive areas or degraded ecosystems. Water availability has been recognized as the most important factor for its development, mainly because of its good adaptability to water stress (Hutmacher et al. 1994). As a result, almond is a crop that responds very positively to deficit irrigation strategies, as has been reported by many authors (Egea et al. 2009, 2010; García et al. 2004; Girona et al. 2005; Romeo and Botía 2006). Among the most notable irrigation strategies is the application of deficit irrigation during the kernel-filling, because of its low impact on yield and fruit quality, with a positive increase on water productivity (Egea et al. 2009). Egea et al. (2010) reported that different strategies of regulated deficit irrigation and partial root-zone drying did not affect the kernel fraction, conforming to the results shown by other authors (Torrecillas et al. 1989; Romero et al. 2004). However, both deficit irrigation strategies promoted significant effects in other productive parameters such as kernel yield, fruit number per tree and kernel weight. Moreover, Egea et al. (2010) have observed a cumulative effect of water stress on kernel weight and crop load. Regarding the fruit-quality parameters, sugar, protein and fat components in fruit are the most reliable components (Nanos et al. 2002; Kodad and Socias 2006). In addition, Egea et al. (2009) demonstrated that severe water stress promoted an increase in kernel-fat components, although other authors, under similar conditions showed contradictory results (Nanos et al. 2002). Kernel-protein content was not found to be significantly affected by deficit irrigation, in contrast with the results of Sánchez et al. (2008). Also, there were no clear effects of deficit irrigation in sugar content. Thus, in arid and semi-arid areas, with reduced water supplies, regulated deficit irrigation and other practices such as partial root-zone drying for almond cultivation is an efficient alternative, in agronomic as well as in economic terms.

Evergreen

Olive (*Olea europea* L.) is the most important evergreen crop in European Mediterranean countries, especially in arid and semi-arid areas, where water is the most limiting factor for agricultural development. This crop has historically been associated with rainfed farming, by its adaptability to long dry periods, with severe water-deficit situations, although, it would promote a decrease in growth and yield (Giménez et al. 1997). This wide adaptability has allowed a very favourable response to irrigation supply, in order to cover the crop water demand during the most critical periods, which has significantly improved production (Moriana et al. 2003; Orgaz and Fereres 2004). According to Baratta et al. (1986), under arid and semi-arid climatic conditions, the olive water demand is around 800–1,000 mm, for the best yield. However, the water resources availability in these areas, coupled with increased water demand from other more productive sectors, have forced the search for different deficit-irrigation strategies to maintain crop viability and ensure a similar benefit when the crop–water demand is completely covered. Therefore, the second phase of fruit growth, when pit hardening is occurring, has been identified as the least sensible to water stress (Goldhamer et al. 1999), whereas the third phase of fruit growth and oil accumulation is highly sensitive to water deficit (Moriana et al. 2003). Tognetti et al. (2005) proposed that olive trees should be irrigated at least with a 66% of ET_c during the third phase of fruit growth, to avoid significantly compromising the yield and fruit quality. Many studies have pointed out the advantages of deficit irrigation in comparison to rainfed practices in olive. D'Andrina et al. (2004) found significant effects of deficit irrigation, improving the final yield and fruit quality with respect to non-irrigated trees. Wahbi et al. (2005) showed that a partial root-zone drying strategy with water supplied of the 50% of ET_c. promoted a lower loss in vegetative growth and crop yield in comparison with fully irrigated trees, although these results were significantly better than for non-irrigated trees.

Finally, many authors have found some advantages from deficit irrigation in terms of fruit quality and oil quality, in comparison to fully irrigated trees (Grattan et al. 2006; Gómez et al. 2005; Muñoz 2005). Patumi et al. (1999) reported a significant reduction in phenolic compounds when a severe water reduction was applied (70% ETc.).

Citrus (*Citrus sinensis* L. Osb) is by far the most important evergreen fruit crop in world trade, and besides, one of the most important fruit crops cultivated in arid and semiarid regions. Orange trees are highly dependent on water, especially in arid and semi-arid zones with low annual rainfall and high evapotranspiration. These conditions promote an accumulated water deficit which requires irrigation. Many authors have reported that the response of the citrus trees to water stress depends mainly on the crop phenology, and the different effects observed are closely related to the timing, duration, crop physiological status, irrigation-water quality, plant genotype, and the degree of stress endured by the crop (Doorenbos and Kassam 1979; Ginestar and Castel 1996; García-Tejero et al. 2008). The negative impact could be mitigated by an improvement in fruit quality, as has been shown by many studies with citrus: verna lemon trees (Sánchez et al. 1989), 'Clementina de Nules' (González and Castel 2000), 'Marisol Clementines' (Verreynne et al. 2001), and sweet orange cv. navelina and salustiana (García-Tejero et al. 2010a, b, c).

In addition, García-Tejero et al. (2010b) found a clear response of fruit yield and morphological and organoleptic fruit characteristics to different regulated deficit-irrigation strategies in citrus (*Citrus sinensis* L. Osb. cv. Navelina) grafted onto carrizo citrange (*Citrus sinensis* L. Osb. x *Poncirus Trifoliata* L. Osb.). The strongest effects were appreciated when a water restriction of 45% of ET_c. were applied during flowering and fruit growth periods (Table 10, treatment C), with an average yield reduction of 21% with regard to fully irrigated trees (100% ET_c). Yield reductions could have been caused by a decline in the fruit number per tree and/or a reduction in fruit weight or fruit size (García-Tejero et al. 2010b). Severe water stress applied during the flowering was reflected in the fruit number per tree (Table 10, treatment A and B) whereas when these water restrictions were applied during the fruit-growth these effects were reflected mainly in fruit size (Table 10, treatment C). However, a greater water supply during fruit growth with respect flowering boosted fruit size, significantly mitigating the reduction during the early developmental stages (Table 10, treatment B).

With regard to the effect of water stress on organoleptic properties, fruit quality was affected mainly in treatments with higher stressed levels during the fruit-

Table 10 Yield and fruit quality under different irrigation strategies at flowering, fruit growth, and maturity stages for citrus trees

Treatment	Yield (kg tree^{-1})	Fruit weight (g)	Equatorial diameter (mm)	TSS (°Brix)	TA
A (55-70-55% ET$_C$)	150.2ab	206.8ab	75.8ab	150.2ab	206.8ab
B (55-70-70% ET$_C$)	149.2ab	218.1b	77.5b	149.2ab	218.1b
C (55-55-70% ET$_C$)	139.7a	182.5a	73.1a	139.7a	182.5a
D (70-70-55% ET$_C$)	166.4ab	206.8ab	76.1ab	166.4ab	206.8ab
Control (100% ET$_C$)	174.3b	233.9b	79.2b	174.3b	233.9b

The values in parenthesis are the irrigation regimes at three phenological stages; *TSS* Total solid soluble, *TA* Titrable acidity

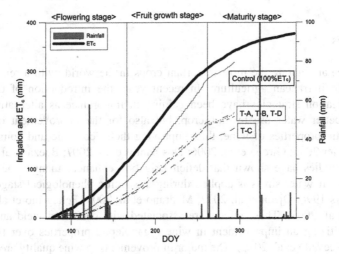

Fig. 12 Cumulative irrigation for each treatment at flowering, fruit growth, and maturity stages for citrus trees

growth and maturity periods (Table 10, treatments A, C and D). Hence, these treatments registered increases in total soluble solids and titrable acidity, coupled with small decreases in maturity index (García-Tejero et al. 2010b). These effects were particularly significant in the most restrictive treatment, in which the severest water restrictions were applied. Noteworthy results were found when severe water stress was applied only during maturity period, with a water supply of 70% of ET$_c$ during flowering and fruit-growth period. In this situation, water stress improves juice quality without lowering the maturity index. In this context, the fruit number registered an average yield reduction of only 5% in relation to the fully irrigated treatment, promoted mainly by a fruit-weight reduction (15% on average; Table 10, treatment D). Figure 12 shows the applied cumulative irrigation for each treatment during the study period.

Thus, according to the results, the deficit-irrigation treatments encourage important water savings without significant impact on yield or the fruit-quality parameters.

On the other hand, González and Castel (2003) observed significant effects on yield, due to a drop in fruit number when the crop underwent moderate to severe water stress. However, this strategy affected neither the fruit weight nor the organoleptic properties. When this reduction was applied during the growing period, the effects were related to fruit weight, but not with significant impact on yield.

García-Tejero et al. (2010a) in a long-term study with sweet orange cv. Salustiano, under different sustainable-deficit irrigation levels (50, 65, 75% ETc.; and a control with 100% ET_c) no significant effect on fruit yield was found. However, differences in some organoleptic fruit properties were found (i.e. total soluble solids, titrable acidity, and peel thickness) in the most restrictive treatments (50 and 65% of ET_c).

Grapevines

Grapevines are among the most important crops in the world, with a remarkable role in Mediterranean agriculture. In recent years the introduction of different deficit-irrigation strategies have been gaining in importance as alternatives not only for better water use by this crop, but also for the improvement of some organoleptic properties of wine, thus improving their economic and commercial value (Keller 2005; Girona et al. 2006; Dos Santos et al. 2007; Baeza et al. 2007).

Many studies have shown that deficit irrigation enhances many wine-quality parameters, if water stress is applied during at specific phenological stages (Dry and Loveys 1998; Ojeda et al. 2002; Medrano et al. 2003; Pellegrino et al. 2005; Chaves et al. 2007). These effects are associated with decreased yield and berry size, prioritizing an improvement in wine organoleptic properties over the final volume (Acevedo et al. 2010). The major improvements in wine quality are related to colour, flavour and aroma as well as the synthesis and concentration of phenological compounds, total soluble solids and anthocyanins (Ojeda et al. 2002; Koundouras et al. 2006). In addition, according to Chaves et al. (2007) the gain in crop water use in deficit irrigation and partial root drying was accompanied by an increase of the $\delta^{13}C$ values in the berries in comparison to fully irrigated plants, suggesting that we can use this methodology to assess the integrated water-use efficiency over the growing season.

Acevedo et al. (2010) demonstrated that some of these properties (i.e. berry diameter, skin:pulp ratio, total and easy extractable anthocyanins) were closely related to midday Ψ_{stem} values. In this sense, Rodrigues et al. (2008) reported that hydraulic feedback and feed-forward root-to-shoot chemical-signalling mechanisms might be involved in the control of stomata in response to decreased soil–water availability, and therefore hydraulic signals play the dominant role.

On other hand, a partial root-zone drying strategy has been applied successfully in commercial grapevines (de la Hera et al. 2007) in order to reduce the vine vigour and water use, maintaining the crop yield and berry size while improving the fruit quality (Dry et al. 2000, 2001). This technique has increased considerably the water-use efficiency in this crop, making it a potential strategy to apply in arid and semi-arid regions. Some field studies, comparing partial root-zone drying and regulated deficit irrigation, with similar amounts of water showed a significant increase in yield with partial root-zone drying practices (Dry et al. 2001) although in other cases positive effects were not found (Intrigliolo et al. 2005; Bravdo et al. 2004; de la Hera et al. 2007).

Annual and Vegetable Crops

Maize

Maize (*Zea mays* L.) is not highly affected by water stress during early vegetative-growth stages, and thus the water demand in this period is low and plants are able to adapt to water stress, reducing the effects on yield (Çakir 2004; Mansouri et al. 2010). In this context, Khang et al. (2000) reported that deficit irrigation can improve the water-use efficiency, and similar findings have been reported by many other authors (e.g. Payero et al. 2006; Farré and Faci 2009; Mansouri et al. 2010). The vegetative and ripening periods are the major periods of water-stress toler-ance. However, for yield similar to that of a fully irrigated crop, it is necessary to supply regular irrigation during the flowering period (Igbadun et al. 2007). Water stress just before anthesis and during silking and seed fill, can promote significant yield reductions (Çakir 2004; Mansouri et al. 2010). Linear relationships have been reported by Mansouri et al. (2010), in an experimental work with four deficit irrigation strategies in this crop. Significant differences were reported in grain yield and kernel weight, these being the most important components to determine the yield variation in deficit-irrigation treatments.

On the other hand, Farré and Faci (2009) reported different results from a field study with deficit irrigation applied during three phenological stages (vegetative, flowering, and grain filling). They concluded that the flowering period was the most critical stage for water deficit, with significant effects on biomass, yield, and harvest index. Yield reduction was promoted mainly by a reduction in the number of grains per square metre. Other treatments based on deficit irrigation strategies applied during vegetative and grain filling did not promote significant effects on yield.

Cotton

Cotton (*Gossypium hirutum* L.) requires an irrigation supply during its develop-ment, thus being a crop with high water demand. The main strategies of deficit irrigation in this crop are based on different crop-water stress situations in its phenological stages. Jalota et al. (2006) found direct linear relationships with high determination coefficients between cotton yield and the amount of water applied. Irrigation restrictions in this crop have increased the focus on improving water-use efficiency. In addition, regulated deficit irrigation has been demonstrated as a good strategy for improving the water-use of efficiency in this crop (Du et al. 2006; Jalota et al. 2006; Suleiman et al. 2007; Singh et al. 2010).

Hutmacher et al. (1995) reported similar yields in various sustainable-deficit irrigation treatments at various growth stages in comparison to the control with fully irrigated plants. Basal et al. (2008) showed that a deficit-irrigation strategy with a 25% of water saving encouraged an increase in water-use efficiency, without significant effects on yield, although, other more severe deficit-irrigation treatments caused several effects on yield and fibre quality. On the other hand, De Tar et al. (1994) pointed out that soil type was more determinant in final yield than was irrigation strategy. Several authors have reported that sustainable-deficit irrigation strategies with high-frequency irrigation can improve the crop yields (Bordovsky and Lyle 1998; Henggeler 1998).

Tomato

Many studies on open-field processing tomato (*Lycopersicon esculentum* M.) grown have demonstrated the relationships between crop-water stress and the influence in crop physiology and productive response (Zegbe et al. 2003; Favati et al. 2009). Nevertheless, the effects of soil–water deficit at different crop stages are not clear at all, being affected mainly by the agro-climatic conditions, varietal crop and deficit-irrigation strategy (Marouelli and Silva 2007). Therefore, Patanè and Cosentino (2010) showed that there was a clear linear relationship between the soil–water deficit and fruit properties (i.e. total solids, soluble solids, titrable acidity, pulp consistency, and fruit size) and yield during the fruit enlargement and ripening.

Other authors such as Marouelli et al. (2004) and Marouelli and Silva (2007) showed that moderate deficit irrigation during the vegetative stage promoted deeper rooting. However, this strategy did not cause significant variations in fruit size, though it allowed water savings of close to 50%. Cahn et al. (2003) reported that water stress during fruit growth and ripening period promoted an increase in total soluble solids and directly affected yield. Johnstone et al. (2005) concluded that moderate deficit irrigation during the fruit ripening and/or maturation growth was a good strategy without affecting the yield.

On the hand, Kirda et al. (2004) pointed out that a partial root-zone drying practice, with a water saving of nearly 50% did not significant alter yield, and had an irrigation water-use efficiency of 56% higher than in the control treatment. In this study they showed that deficit irrigation was more efficient than sustainable-*deficit* irrigation, with similar amounts of irrigation water applied. Other parameters affected by deficit irrigation were fruit weight, fruit size, and total solids, although there were no significant differences between partial root-zone drying (50%) and control treatment.

Potato

Potato (*Solanum tuberosum* L.) has been recognized as moderately sensitive to deficit irrigation, needing adequate water from tuber initiation to near maturity, and it is highly influenced by the amount and timing of irrigation water applied.

For these reasons, it is necessary to identify the most critical growth stages for applying a deficit-irrigation strategy. Liu et al. (2006) showed that yield and tuber size are considerably affected by water stress. These authors noted that an adequate amount of water during the flowering and yield-formation stages with a moderate water stress during vegetative and maturity stages produced optimum yield. On other hand, Iqbal et al. (1999) showed that the timing of water stress promoted several effects in tuber yield. Consequently, a crop water stress during the ripening had no significant impact on yield, whereas water stress during the first pheno-logical stages caused the greatest yield reduction, these being the most critical periods for applying a deficit-irrigation strategy. Kashyap and Panda (2003), in a field study with much sustainable-deficit irrigation found good linear relations between the water amount applied and some yield parameters, such as fresh tuber yield, dry tuber yield, plant dry matter, and total dry matter. They reported that when the crop evapotranspiration decreased, the fresh tuber yield also decreased, with a highly significant linear correlation.

Liu et al. (2006) found no significant differences with respect to other tradi-tional practices of deficit irrigation (sustainable-deficit irrigation, with similar amounts of water applied), although other authors emphasize the advantages of deficit irrigation in this crop, with significant water savings (closely to 30–50%), maintaining the yield and improving the water-use efficiency by 60% (Shahnazari et al. 2007).

Deficit irrigation strategy has been defined as a good alternative for optimising water use in agriculture, keeping in many cases the viability of agro-ecosystems as well as in some cases improving certain organoleptic properties of the final product and offering substantial irrigation-water savings.

Moreover, most researchers emphasize the need to further develop studies for the different agro-climatic conditions in which crops are developed. The estab-lishment of either deficit-irrigation strategy should be considered one of the main elements of the study area, and preliminary results have been obtained for similar conditions.

In most cases, the use of such strategies can significantly raise water produc-tivity, yielding significant results in terms of financial income for the farmer.

Finally, in addition to such strategies, the introduction of new tools for more efficient management of the water resources are a viable option to improve both the water balance and productivity, an aspect that is crucial for promoting sus-tainable irrigated agriculture.

6 Conclusions

With the expanding and more affluent population throughout the world, demand for agricultural products will increase rapidly over the coming decades, with serious implications for agricultural water demand. Symptoms of water scarcity are increasingly threatening ecosystem services and the sustainability of food production. Agriculture is the dominant user of flowing water, rainfall water, and land, and as the economic sector traditionally most closely related to a wide range of social and environmental issues that touch on human behaviour and involve hard policy decisions. It is well-known that agriculture has a long tradition and is adapted to produce under conditions of climatological and economic risk. Also, over the years agriculture has suffered from changing policies and been guided by changing and conflicting paradigms ranging from high input and unmeasured exploitation of natural resources for increased production. More recently, a focus is being placed on food security with the current emphasis on crop production and market liberalisation towards a global agricultural sector, accompanied by growing concerns for the ecological systems and the biodiversity values.

The main challenge of water-management policies will be to protect and pre-serve water, a highly sensitive and indispensable natural resource. To achieve this, policy makers must work with the principles of flexibility and adaptive manage-ment in response to changes in society's expectations, development needs, and environmental quality. A technical standard for pollution that may appear adequate at one point in time may require tightening and strengthening in light of new scientific evidence or ecological developments, such as droughts.

In irrigated agriculture, water-use efficiency is broader in scope than most agronomic applications and must be considered on a basin, watershed, catchment, or irrigation district scale.

The ever-increasing demand of irrigation water, together with the growing difficulties and costs for developing new resources, makes it necessary to carry out field surveys aiming at a better management of irrigation systems as well as to evaluate specific operational and management decisions. Current recommenda-tions of irrigation management service should be valid in both an economic and technical sense. For the short term, efforts to conserve water through quantity restrictions on surface water can greatly affect farm profits and crop-water pro-duction functions. For the long term, water quantity restrictions could encourage more efficient irrigation systems or practices. Therefore, future perspectives of efficient irrigation systems should take into account the following matters:

1. The meeting of future challenges posed by food security, i.e. raising production while conserving natural resources.
2. Sustainable water management in agrosystems to improve food quality and safety.
3. Rehabilitation and modernization of the already existing irrigation systems and design of new large-scale irrigation schemes.
4. Application of new technologies with the scope of improving the management, maintenance, and operation of irrigation systems.
5. Development of crop models for optimising irrigation and new deficit-irrigation strategies that would decrease the effects of water shortage.
6. Investigation and technological development in "irrigation scheduling" using indicators of plant water-stress status.

7. Study of water relations and ecophysiological behaviour of species adapted to drought environments and plant breeding for improved drought tolerance under climate-change scenarios.
8. The development of a comprehensive approach that integrates all these factors into irrigation-project selection, requiring further research on the processes governing climate changes, the impact that increased atmospheric CO_2 will exert on vegetation and runoff, the effect of climate variables on crop-water requirements and the impact of climate on infrastructure performance.
9. Improvement of water-management techniques to protect exploitable resources and to develop more efficient supply systems starting from water pumping until the delivery to end-users. Such techniques will provide integrated tools towards the improvement of water-use efficiency and preserving a rational equilibrium between supply and demand.

Different scenarios have been developed to explore a number of issues, such as the expansion of irrigated agriculture, massive increases in food production from rainfed lands and water productivity trends. Many researchers consider the rainfed agriculture as risky, but it has the potential to produce large amounts of cereal in dry regions. For this potential to be realized, farmers, researchers and policy-makers must work together to improve technology and reconsider economic policies.

Sustainable agricultural land and water-management practices are essential to boost productivity, promote regional growth, and protect the environment. This entails addressing the potentially negative impacts of land and water management, such as water logging and salinity and the diverse types of impact on ecosystem health and biodiversity.

Without appropriate management, irrigated agriculture can be detrimental to the environment and endanger sustainability. Irrigated agriculture is facing growing competition for low-cost, high-quality water.

Acknowledgments Partly of the research work leading to this publication was sponsored by the following research projects "Hydrological and erosive processes, and biomass assessment and organic carbon sequestering under different land uses in the Mediterranean agrarian watershed El Salado, Lanjaron" (SE Spain) (RTA2007-00008-00-00) and "Strategies for the improvement irrigation management under climatic change. Integration of modelling techniques and deficit irrigation strategies" (RTA2008-00006-CO2-02) both granted by INIA, Spain, and cofinanced by FEDER funds (European Union).

References

Abdullah K (2006) Use of water and land for food security and environmental sustainability. Irrig Drain 55:219–222
Aberra Y (2004) Problems of the solution: intervention into small-scale irrigation for drought proofing in the Mekele Plateau of northern Ethiopia. Geogr J 170:226–237

Aboudrare A, Debaeke P, Bouaziz A, Chekli H (2006) Effects of soil tillage and fallow management on soil water storage and sunflower production in a semi-arid Mediterranean climate. Agric Water Manag 83:183–196

Abu-Zeid MA (1998) Water and sustainable development: the vision for world water, life and the environment. Water Policy 1:9–19

Acevedo OCS, Ortega F, Fuentes F (2010) Effects of grapevine (Vitis vinifera L.) water status on water consumption, vegetative growth and grape quality: An irrigation scheduling application to achieve regulated deficit irrigation. Agric Water Manag 97:956–964

Aggarwal PK (2003) Impact of climate change on Indian agriculture. J Plant Biol 30:189–198

Alarcón JJ, Domingo R, Green SR, Sánchez-Blanco MJ, Rodríguez P, Torrecillas A (2000) Sap flor as indicador of transpiration and the water status of young apricot trees. Plant Soil 227:77–85

Alcamo J, Döll P, Henrichs T, Kaspar F, Lehner B, Rösch T, Siebert S (2003) Global estimates of water withdrawals and availability under current and future "business-as-usual" conditions. Hydrol Sci J 48:339–348

Alexandridis TK, Zalidis GC, Silleos NG (2008) Mapping irrigated area in Mediterranean basins using low cost satellite Earth Observation. Comput Electron Agric 64:93–103

Ali MH, Talukder MSU (2008) Increasing water productivity in crop production—a synthesis. Agric Water Manag 95:1201–1213

Ali MH, Hoque MR, Hassan AA, Khair A (2007) Effects of deficit irrigation on yield, water productivity, and economic returns of wheat. Agric Water Manag 92:151–161

Allen RG, Pruitt WO (1986) Rational use of FAO Blaney-Criddle formula. J Irrig Drain Eng 112:139–155

Allen RG, Pereira LS, Raes D, Smith M (1998) Crop evapotranspiration: guidelines for computing crop water requirements. Irrigation and Drainage Paper No. 56, FAO, Rome, Italy, p 300

Anapalli SS, Ahuja LR, Ma L, Timlin DJ, Stockle CO, Boote KJ, Hoogenboom G (2008) Current water deficit stress simulations in selected agricultural system simulation models. In: Ahuja LR, Reddy VR, Saseendran SA, Yu Q (eds), Response of crops to limited water: understanding and modeling water stress effects on plant growth processes. Madison, WI, Am Soc Agr, Crop Sci Soc Am, Soil Sci Soc Am pp 1–38

Appelgren B (2004) Water and ethics Water in Agriculture, Essay 5. UNESCO, Paris, France, p 40

AQUASTAT (2005) FAO-database. http://www.fao.org/ag/agl/aglw/aquastat/main/index.stm

Araus JL (2004) The problems of sustainable water use in the Mediterranean and research requirements for agriculture. Ann Appl Biol 144:259–272

Araus JL, Bort J, Steduto P, Villegas D, Royo C (2003) Breeding cereals for Mediterranean conditions: ecophysiological clues for biotechnology application. Ann Appl Biol 142:129–141

Ashton PJ (2002) Avoiding conflicts over Africa's water resources. Ambio 31:236–242

Bacon MA (2004) Water use efficiency in plant biology. Blackwell publishing, Oxford, p 327

Baeza P, Sánchez DMP, Centeno A, Junquera P, Linares R, Lissarrague JR (2007) Water relations between leaf water potential, photosynthesis and agronomic vine response as a tool for establishing thresholds in irrigation scheduling. Sci Hortic 114:151–158

Bahri A (2002) Integrated management of limited water resources in Tunisia. Research topic priorities, In: Rodriguez R (ed) Identifying priority tools for cooperation. INCO-MED Workshops, European Commission, Brussels, Belgium

Bangoura S (2002) Water harvesting techniques in West and Central Africa. In: Dupuy A, Lee C, Schaaf T (eds) Proceedings of the international seminar on combating desertification: freshwater resources and the rehabilitation of degraded areas in the Drylands, Samantha Wauchope. Fushia publishing Paris, N'Djamena Chad, pp 20–26

Baratta B, Caruso T, Di Marco L, Inglese P (1986) Effects of irrigation on characteristics of olives in 'Nocellara del Belice' variety. Olea 17:195–198

Barron J (2004) Dry spell mitigation to upgrade semi-arid rainfed agriculture. Water harvesting and soil nutrient management for smallholder maize cultivation in Machakos, Kenya. Doctoral thesis in Natural Resource Management, Department of Systems Ecology, Stockholm University, Stockholm, pp 1–39. http://www.diva-portal.org/su/theses/abstract.xsql?dbid=98

Bartolini F, Bazzani GM, Gallerani V, Raggi M, Viaggi D (2007) The impact of water and agriculture policy scenarios on irrigated farming systems in Italy: An analysis based on farm level multi-attribute linear programming models. Agric Syst 93:90–114

Basal H, Dagdelen N, Unay A, Yilmaz E (2008) Effects of deficit drip irrigation ratios on cotton (*Gossypium hirsutum L.*) yield and fibre quality. J Agron Crop Sci 195:19–29

Batchelor C, Singh A, Rao RM, Butterworth J (2002) Mitigating the potential unintended impacts of water harvesting. In: Paper presented at the IWRA international regional symposium 'water for human survival'. New Delhi, India

Batty JC, Keller J (1980) Energy requirements for irrigation. In: Pimentel D (ed) Handbook of energy utilization in agriculture. CRC Press, Boca Raton, pp 35–44

Behboudian MH, Lawes GS, Griffiths KM (1994) Influence of water deficit on water relations photosynthesis and fruit growth in Asian pear (Pyrus serotina Rehd). Sci Hortic 60:89–99

Behboudian MH, Dixson J, Pothamshetty K (1998) Plant and fruit response of lysimeter-grown 'Braeburn' apple to deficit irrigation. J Hort Sci Biotech 73:781–785

Bekele S, Tilahun K (2007) Regulated deficit irrigation scheduling of onion in a semiarid region of Ethiopia. Agric Water Manag 89:148–152

Benham B, Schneekloth J, Elmore R, Eisenhauer D, Specth J (1999) Irrigating Soybeans. Cooperative extension, Institute of Agriculture and Natural Resources, University of Nebraska, Lincoln, NE. http://www.ianr.unl.edu/pubs/fieldcrops/g1367.htm

Bennett EM, Carpenter SR, Caraco NF (2001) Human impact on erodable phosphorus and eutrophication: a global prospective. Bio Sci 51:227–234

Bennie ATP, Hensley M (2001) Maximizing precipitation utilization in dryland agriculture in South Africa—a review. J Hydrol 241:124–139

Bergez JE, Deumier JM, Lacroix B, Leroy P, Wallach D (2002) Improving irrigation schedules by using a biophysical and a decisional model. Eur J Agron 16:123–135

Bhattarai M, Barker R, Narayanamoorthy A (2007) Who benefits from irrigation development in India? Implication of irrigation multipliers for irrigation financing. Irrig Drain 56:207–225

Bilskie J (1997) Using dielectric properties to measure soil water content. Sensors Mag 14:26–32

Blaney HF, Criddle WD (1950) Determining water needs from climatological data. U S D A Soil Conservation Service. SOS – TP, USA, pp 8–9

Blavet D, De Noni G, Le Bissonnais Y, Leonard M, Maillo L, Laurent JY, Asseline J, Leprun JC, Arshad MA, Roose E (2009) Effect of land use and management on the early stages of soil water erosion in French Mediterranean vineyards. Soil Till Res 106:124–136

Blum A (2009) Effective use of water (EUW) and not water-use efficiency (WUE) is the target of crop yield improvement under drought stress. Field Crops Res 112:119–123

Bodnar F, de Graaff J (2003) Factors influencing adoption of soil and water conservation in southern Mali. Land Degrad Develop 14:515–525

Boland AM, Corrie JA, Bewsell D, Jerie PH (1998) Best management practice and benchmarking for irrigation, salinity and nutrients of stone and pome fruit. In: Proceedings of IAA conference. Brisbane, Australia

Boland AM, Corrie JA, Bewsell D, Jerie PH (2001) Final Report for Project I7044—development of benchmarks and best management practices (BMP's) for perennial horticulture, Murray Darling Basin Commission, Irrigation strategic investigation and education program. Australia

Boland AM, Zhierl A, Beaumont J (2002) Guide to best practice in water management—orchard crops. Department of Primary Industries, Victoria Australia

Boland AM, Bewsell D, Kaine G (2006) Adoption of sustainable irrigation management practices by stone and pome fruit growers in the Goulburn/Murray Valleys, Australia. Irrig Sci 24:137–145

Bordovsky JP, Lyle WM (1998) Cotton irrigation with LEPA and subsurface drip irrigation systems on the Southern High Plains. In: Proceedings of beltwide cotton conference, Nat Cotton Council, Memphis, TN, USA, pp 409–412

Bos MG (1980) Irrigation efficiencies at crop production level. ICID Bull 29:18–25

Bos MG (1985) Summary of ICID definitions of irrigation efficiency. ICID Bull 34:28–31

Botha JJ, van Rensburg LD, Botha JJ, Anderson JJ, Groenewald DC, Kundhlande G, Baiphethi MN, Viljoen MF (2004) Evaluating the sustainability of the in-field rainwater harvesting crop production system. In: ICID-FAO international workshop on water harvesting and sustainable agriculture. Moscow, Russia, pp 1–13

Bouwer H (2002) Integrated water management for the 21st century: problems and solutions. J Irrig Drain Eng 128:193–202

Bowonder B, Ramana KV, Rajagopal R (1986) Waterlogging in irrigation projects. Sadhana 9:177–190

Brakke TW, Verma SB, Rosenberg NJ (1978) Local and regional components of sensible heat advection. J Appl Meteorol 17:955–963

Brandelik A, Hübner C (1996) Soil moisture determination—accurate, large and deep. Phys Chem Earth 21:157–160

Bravdo B, Naor A, Zahavi T, Gal Y (2004) The effects of water stress applied alternately to part of the wetting zone along the season (PRD-partial root-zone drying) on wine quality, yield and water relations of red wine grapes. Acta Hortic 664:101–109

Bray E (1997) Plant responses to water deficit. Trends Plant Sci 2:48–54

Breeuwsma A, Silva S (1992) Phosphorous fertilization and environmental effects in The Netherlands and Po Region (Italy). Winand Staring Centre (SC-DLO) Report 57, Wageningen, The Netherlands

Broner I (2002) Irrigation scheduling: the water balance approach. Colorado State University Cooperative Extension, Bulletin No. 4. 707, CO, USA

Brown LR (1999) Feeding nine billion. In: Starke L (ed) State of the world 1999. W.W. Norton and Co., New York, pp 115–132

Brown LR (2002) Water deficits growing in many countries: water shortages may cause food shortages. Earth Policy Institute, Washington DC. http://www.earth-policy.org/Updates/Update15.htm

Bruinsma J (2003) World agriculture: towards 2015/2030, An FAO perspective. Earthscan Publications, London

Brumbelow K, Georgakakos A (2001) An assessment of irrigation needs and crop yield for the United States under potential climate changes. J Geophys Res Atmos 106:27383–27405

Burke S, Mulligan M, Thornes JB (1999) Optimal irrigation efficiency for maximum plant productivity and minimum water loss. Agric Water Manag 40:377–391

Burt CM, Clemmens AJ, Strelkoff TS, Solomon KH, Bliesner RD, Hardy L, Howell TA, Eisenhauer DE (1997) Irrigation performance measures: efficiency and uniformity. J Irrig Drain Eng 123:423–442

Cahn MD, Herrero EV, Hanson BR, Snyder RL, Hartz TK, Miyao EM (2003) Effects of irrigation cut-off on processing tomato fruit quality. Acta Hortic 613:75–80

Çakir R (2004) Effect of water stress at different development stages on vegetative and reproductive growth of maize. Field Crops Res 89:1–16

Calzadilla A, Rehdanz K, Tol RSJ (2010) The economic impact of more sustainable water use in agriculture: a computable general equilibrium analysis. J Hydrol 384:292–305

Campbell GS (1991) An overview of methods for measuring sap flow in plants. In: Collected summaries of papers at the 83rd annual meeting of the American Society of Agronomy, Division A-3: Agroclimatology and agronomic modelling, Denver, Colorado, USA, pp 2–3

Carpenter SR, Caraco NF, Correll DL, Howarth RW, Sharpley AN, Smith VH (1998) Nonpoint pollution of surface waters with phosphorus and nitrogen. Ecol Appl 8:559–568

Caspari HW, Behboudian MH, Chalmers DJ, Clothier BE, Lenz F (1996) Fruit characteristics of 'Hosui' Asian pear under deficit irrigation. HortScience 31:162

Cayci G, Heng LK, Öztürk HS, Sürek D, Kütük C, Sağlam M (2009) Crop yield and water use efficiency in semi-arid region of Turkey. Soil Till Res 103:65–72

CEC (1994) Desertification and land degradation in the European Mediterranean. Report EUR 14850, Commission of the European Communities, Directorate-General XII, Brussels, Belgium

Čermák J (1995) Methods for studies of water transport in trees, especially the stem heat balance and scaling. In: Proceedings of the 32th course in Applied Ecology, San Vito di Cadore, University of Padova, Italy, pp 4–8

Chaves MM, Santos TP, Souza CR (2007) Deficit irrigation in grapevine improves water-use efficiency while controlling vigour and production quality. Ann Appl Biol 150:237–252

Cheesman JM (1991) Effects of salinity on stomatal conductance, photosynthetic capacity and carbon isotope discrimination of salt tolerant (Gossypium hirsutum L.) and salt sensitive (Phaseolus vulgaris L.) C3 nonhalophytes. Plant Physiol 95:628–635

Chen DX, Coughenour MB (2004) Photosynthesis, transpiration, and primary productivity: scaling up from leaves to canopies and regions using process models and remotely sensed data. Glob Biogeochem Cycl 18:1–15

Chenoweth J (2008) Minimum water requirement for social and economic development. Desalination 229:245–256

Chiotti Q, Johnston T, Smit B, Ebel B (1997) Agricultural response to climate change: a preliminary investigation of farm-level adaptation in Southern Alberta. In: Ilbery B, Chiotti Q, Rickard T (eds.), Agricultural restructuring and sustainability, CAB International, pp 201–218

Cohen Y (1993) Thermoelectric methods for measurement of sap flow in plants. In: Standill G (ed) Advances in bioclimatology. Springer, Berlin, pp 63–89

Costanza R (1995) Economic growth, carrying capacity, and the environment. Ecol Econ 15:89–90

Covich AP (1993) Water and ecosystems. In: Gleick PH (ed) Water in Crisis. Oxford University Press, New York, pp 40–55

Cowan IR (1982) Regulation of water use in relation to carbon gain in higher plants. In: Lange OL, Nobel PS, Osmond CB, Ziegler H (eds) Physiological Plant Ecology. II. Water relations and carbon assimilation. Encyclopedia of Plant Physiology (N.S.), vol 12B. Springer, Berlin, pp 589–613

Critchley W, Siegert KC (1991) Water harvesting: a manual for the design and construction of water harvesting schemes for plant production. Food Agric Organ, Rome

Cui N, Du T, Kang S, Li F, Zhang J, Wang M, Li Z (2008) Regulated deficit irrigation imporved fruit quality and water use efficncy of pear-jujube trees. Agric Water Manag 95:698–706

Cuttle SP, Macleod JCA, Chadwick DR, Scholefield D, Haygarth PM, Newell PP, Harris D, Shepherd MA, Chambers BJ, Humphrey R (2007) An inventory of methods to control diffuse water pollution from agriculture, DWPA, User manual ES0203. Defra, London

D'Andrina R, Lavini A, Morelli G, Patumi M, Terenziani S, Calandrelli D, Fragnito F (2004) Effects of water regimes of five picking and duble aptitude olive cultivars (Olea europea L.). J Hort Sci Biotechnol 79:18–25

Davies WJ, Zhang J (1991) Root signals and the regulation of growth and development of plant in drying soil. Ann Rev Plant Physiol Plant Mol Biol 42:55–76

Day JC, Hughes DW, Butcher WR (1992) Soil, water and crop management alternatives in rainfed agriculture in the Sahel: an economic analysis. Agric Econ 7:267–287

De Boer J (1993) Building sustainable agricultural systems: economic and policy dimensions. Development studies paper series. Winrock International Institute for Agricultural Development, Morrilton

De Fraiture C, Wichelns D (2010) Satisfying future water demands for agriculture. Agric Water Manag 97:502–511

De Fraiture C, Molden D, Wichelns D (2010) Investing in water for food, ecosystems, and livelihoods: An overview of the comprehensive assessment of water management in agriculture. Agric Water Manag 97:495–501

De Graaff J (2000) Land–water linkages in rural watersheds. Electronic Workshop, Background Paper 5, FAO, Rome, Italy. http://www.fao.org/AG/agL/watershed/watershed/papers/paperbck/papbcken/degraff.pdf

De Groot RBA, Hermans LM (2009) Broadening the picture: negotiating payment schemes for water-related environmental services in the Netherlands. Ecol Econ 68:2760–2767

De la Hera ML, Romero P, Gómez-Plaza E, Martínez A (2007) Is partial root-zone drying an effective irrigation technique to improve water use efficiency and fruit quality in field-grown wine grapes under semiarid conditions? Agric Water Manag 87:261–274

De Swaef T, Steppe K, Lemeur R (2009) Determining reference values for stem water potential and maximum daily shrinkage in young apple trees based on plant responses to water deficit. Agric Water Manag 96:541–550

De Tar WR, Phene CJ, Clark DA (1994) Subsurface drip vs. furrow irrigation: 4 years of continuous cotton on sandy soil. In: Proceedings of the beltwide cotton conference, Memphis, Tennessee: National Cotton Council, pp 542–545

Defra (2007) Observatory programme indicators: indicator DA3—nitrate and phosphorate. http://statistics.defra.gov.uk/esg/ace/da3_fact.htm. Accessed July 2007

Del-Pozo A, Pérez P, Morcuende R, Alonso A, Martínez CR (2005) Acclimatory responses of stomatal conductance and photosynthesis to elevated CO_2 and temperature in wheat crops grown at varying levels of N supply in a Mediterranean environment. Plant Sci 169:908–916

Deng XP, Shan L, Zhang H, Turner NC (2006) Improving agricultural water use efficiency in arid and semiarid areas of China. Agric Water Manag 80:23–40

Desborough CE, Pitman AJ, Irannejad P (1996) Analysis of the relationship between bare soil evaporation and soil moisture simulated by 13 land surface schemes for a simple non-vegetated site. Global Planet Change 13:47–56

Di Paolo E, Rinaldi M (2008) Yield response of corn to irrigation and nitrogen fertilization in a Mediterranean environment. Field Crop Res 105:202–210

Dinar A, Howitt RE (1997) Mechanisms for allocation of environmental control cost: empirical tests of acceptability and stability. J Environ Manag 42:183–203

Dinar A, Xepapadeas A (1998) Regulating water quantity and quality in irrigated agriculture. J Environ Manag 54:273–289

Dinar A, Zilberman D (1991) The economics of resource-conservation, pollution-reduction technology selection: the case of irrigation water. Resour Energy 13:323–348

Dinar A, Rhoades JD, Nash P, Waggoner BL (1991) Production functions relating crop yield, water quality and quantity, soil salinity and drainage volume. Agric Water Manag 19:51–66

Dinar A, Campbell MB, Zilberman D (1992) Adoption of improved irrigation and drainage reduction technologies under limiting environmental conditions. Environ Resour Econ 2:373–398

Döll P (2002) Impact of climate change and variability on irrigation requirements: a global perspective. Clim Change 54:269–293

Domagalski JL, Dubrovsky NM (1992) Pesticide residues in ground water of the San Joaquin Valley, California. J Hydrol 130:299–338

Doorenbos J, Kassam A (1979) Yield response to water. FAO Irrigation and Drainage 33. Rome, Italy

Doorenbos J, Pruitt WO (1977) Crop water requirements. FAO Irrigation and Drainage paper no. 24. FAO Rome

Doran JW, Sarrantonio M, Liebig MA (1996) Soil health and sustainability. Adv Agron 56:1–54

Dos Santos TP, Lopes CM, Rodrigues ML, De Souza CR, Ricardo DSJM, Maroco JO, Pereira JS, Chaves MM (2007) Effects of deficit irrigation strategies on cluster microclimate for improving fruit composition of Moscatel field-grown grapevines. Sci Hortic 112:321–330

Downward SR, Taylor R (2007) An assessment of Spain's Programa AGUA and its implications for sustainable water management in the province of Almería, Southeast Spain. J Environ Manag 82:277–289

Dregne H, Chou NT (1992) Global desertification dimensions and costs. In: Dregne H (ed) Degradation and restoration of Arid Lands. Texas Tech University, Lubbock, pp 249–282

Dry P, Loveys BR (1998) Factors influencing grapevine vigour and the potential for control with partial rootzone drying. Austr J Grape Wine Res 4:140–148

Dry PR, Loveys BR, Stoll M, Stewart D, McCarthy MG (2000) Partial root-zone drying—an update. Aust Grapegrower Winemaker 438:35–39

Dry PR, Loveys BR, McCarthy MG, Stoll M (2001) Strategic management in Australian vineyards. J Int Sci Vigne Vin 35:1–11

Du T, Kang S, Zhang J, Li F, Hu X (2006) Yield and physiological responses of cotton to partial root-zone irrigation in the oasis field of northwest China. Agric Water Manag 84:41–52

Duchemin M, Hogue R (2009) Reduction in agricultural non-point source pollution in the first year following establishment of an integrated grass/tree filter strip system in southern Quebec (Canada). Agric Ecosyst Environ 131:85–97

Dudgeon D (2000) Large-scale hydrological changes in tropical Asia: prospects for riverine biodiversity. BioScience 50:793–806

Durán ZVH, Francia MJR, Rodríguez PCR, Martínez RA, Cárceles RB (2006) Soil-erosion and runoff prevention by plant covers in a mountainous area (SE Spain): Implications for sustainable agriculture. Environmentalist 26:309–319

Durán ZVH, Rodríguez PCR, Francia MJR, Martínez RA, Arroyo PL, Cárceles RB (2008) Benefits of plant strips for sustainable mountain agriculture. Agron Sustain Develop 28:497–505

Durán ZVH, Rodríguez PCR, Martínez RA, Francia MJR, Cárceles RB (2009a) Measures against soil erosion in rainfed olive orchards on slopes (SE Spain): impact of plant strips on soil-water dynamics. Pedosphere 4:453–464

Durán ZVH, Rodríguez PCR, Flanagan CD, Martínez RA, Francia MJR (2009b) Agricultural runoff: new research and trends. In: Hudspeth ChA, Reeve TE (eds) Agricultural runoff coastal engineering and flooding. Nova Science Publishers, Hauppauge, pp 27–48

Durán ZVH, García-Tejero I, Francia MJR, Cárceles RB, Talavera RM, Muriel FJL (2010) Soil erosion: causes, processes and effects. In: Columbus F (ed) Soil erosion, Nova Science Publishers, Hauppauge, pp 1–36

EA (2007a) Environment Agency, Environmental facts and figures: nitrates in rivers and groundwater. http://www.environmentagency.gov.uk/yourenv/eff1190084/water/210440/210566/?lang=

EA (2007b) Environment Agency. Environmental indicators: nutrients in rivers background and data. http://environment.gov.uk/yourenv/432430/432434/432 487/438449/438463/?lang

Earth Policy Institute (2002) Rising temperatures and falling water tables raising food prices. http://www.earth-policy.org/Updates/Update16.html

Easter WK (1993) Economic failure plagues developing countries public irrigation: an assurance problem. Water Resour Res 29:1913–1922

Ebel RC, Proebsting EL, Patterson ME (1993) Regulated deficit irrigation may alter apple maturity, quality, and storage life. HortScience 28:141–143

Ebel RC, Proebsting EL, Evans RG (2001) Apple tree and fruit responses to early termination of irrigation in a semi-arid environment. HortScience 36:1197–1201

EEA (1995) The Dobris assessment. In: Stanners D, Boureau P (eds.), Europe's environment, Office for official publications of the European communities, Luxemburg

Egea G, González RMM, Baille A, Nortes PA, Sánchez BP, Domingo R (2009) The effects of contrasted deficit irrigation strategies on the fruit growth and kernel quality of mature almond trees. Agric Water Manag 96:1605–1614

Egea G, Nortes PA, González RMM, Baille A, Domingo R (2010) Agronomic response and water productivity of almond trees under contrasted deficit irrigation regimes. Agric Water Manag 97:171–181

Ellis JJ, Tengberg A (2000) The impact of indigenous soil and water conservation practices on soil productivity: examples from Kenya, Tanzania and Uganda. Land Degrad Develop 11:19–36

Enciso JM, Unruh BL, Colaizzi PD, Multer WL (2003) Cotton response to subsurface drip irrigation frequency under deficit irrigation. Appl Eng Agric 19:555–558

Engelman R, LeRoy P (1993) Conserving land: population and sustainable food production. Population Action International, Washington DC

English M, Raja SN (1996) Perspectives on deficit irrigation. Agric Water Manag 32:1–14

Estienne P, Godard E (1970) Climatologie. A, Collin Editino, p 365

Evett SR (1998) Coaxial multiplexer for time domain reflectometry measurement of soil water content and bula electrical conductivity. Trans ASAE 42:361–369

Failla O, Zocchi Z, Treccani C, Socucci S (1992) Growth, development and mineral content of apple fruit in different water status conditions. J Hort Sci 67:265–271

Falkenmark M, Rockström J (2004) Balancing water for man and nature: the new approach to eco-hydrology. EarthScan, London

Falkenmark M, Fox P, Persson G, Rockström J (2001) Water harvesting for upgrading of rainfed agriculture, problem analysis and research needs. SIWI Report 11, Stockholm Environmental Institute

Fan T, Stewart BA, Yong W, Junjie L, Guangye Z (2005a) Long-term fertilization effects on grain yield, water-use efficiency and soil fertility in the dryland of Loess Plateau in China. Agric Ecosyst Environ 106:313–329

Fan T, Stewart BA, Payne WA, Wang Y, Song S, Luo J, Robinson CA (2005b) Supplemental irrigation and water: yield relationships for plasticulture crops in the loess plateau of China. Agron J 97:177–188

FAO (2000) Water and agriculture in the Nile Basin. Nile Basin initiative Report to ICCON, FAO/AGL/29/2000. FAO, Rome, Italy

FAO (2002a) Agriculture: towards 2015/30. Technical Interim Report, FAO, Rome, Italy. http://www.fao.org/es/esd/at2015/toc-e.htm

FAO (2002b) Crops and drops: making the best use of water for agriculture. Food and Agriculture Organization, United Nations, Rome

FAO (2003) World agriculture: towards 2015/2030. An FAO perspective. FAO/Earthscan, Rome

FAO (2005a) Food and Agriculture Organization of the United Nations: FAO Statistical Databases (FAOSTAT). http://faostat.fao.org/

FAO (2005b) Food and Agriculture Organization of the United Nations: review of agricultural water use per country, Rome, Italy. http://www.fao.org/ag/agl/aglw/aquastat/wateruse/index.stm

FAO (2007) Food and Agriculture Organization, FAOSTAT database. http://faostat.fao.org/

FAO (2008). Database. Food and Agriculture Organization, Rome, Italy. http://faostat.fao.org/

Farquhar GD, Sharkey TD (1982) Stomatal conductance and photosynthesis. Annu Rev Plant Physiol 33:317–345

Farré I, Faci JM (2006) Comparative response of maize (Zea mays L.) and sorghum (Sorghum bicolour L. Moench) to deficit irrigation in a Mediterranean environment. Agric Water Manag 83:135–143

Farré I, Faci JM (2009) Deficit irrigation in maize for reducing agricultural water use in a Mediterranean environment. Agric Water Manag 96:383–394

Fatondji D (2002) Organic fertilizer decomposition, nutrient release and nutrient uptake by millet crop in a traditional land rehabilitation techniques (Zai), in the Sahel. In: ZEF Ecology and Development Series No. 1. Cuvillier Verlag, Bonn

Faures JM, Hoogeveen J, Bruinsma J (2002) The FAO irrigated area forecast for 2030. FAO, Rome

Favati F, Lovelli S, Galgano F, Miccolis V, Di Tommaso T, Candido V (2009) Processing tomato quality as affected by irrigation scheduling. Sci Hortic 122:562–571

Fereres E (1984) Adaptation des vegetaux a la sechersse. Strategies et mecanismes, Bull Soc Bot Fr 131, Actual Bot 1, pp 17–37

Fereres E (1996) Irrigation scheduling and its impact on the 21st century. In: Camp C, Sadler E, Yoder R (eds) Evapotranspiration and Irrigation Scheduling. A.S.A.E. San Antonio, Texas, pp 547–553

Fereres E, Goldhamer DA (2003) Suitability of stem diameter variations and water potential as indicators for irrigation scheduling of almond trees. J Hort Sci Biotech 78:139–144

Fereres E, Soriano MA (2007) Deficit irrigation for reducing agricultural water use. J Exp Bot 58:147–159

Fernández JE, Cuevas MV (2010) Irrigation scheduling from stem diameter variations: a review. Agric Forest Meteorol 150:135–151

Fernández JE, Palomo MJ, Díaz-Espejo A, Clothier BE, Green SR, Girón IF, Moreno F (2001) Heat-pulse measurements of sap flow in olives for automating irrigation: tests, root flow and diagnostics of water stress. Agric Water Manag 51:99–123

Fernández JE, Green SR, Caspari HW, Diaz EA, Cuevas MV (2008a) The use of sap flow measurements for scheduling irrigation in olive, apple and Asian pear trees and in grapevines. Plant Soil 305:91–104

Fernández JE, Romero R, Montaño JC, Diaz EA, Muriel FJL, Cuevas MV, Moreno F, Girón IF, Palomo MJ (2008b) Design and testing of an automatic irrigation controller for fruit tree orchards, based on sap flow measurements. Aust J Agric Res 59:589–598

Ferré PA, Topp GC (2002) Time domain reflectometry. In: Dana JH, Topp GC (eds) Methods of soil analysis, Part 4-physical methods. American Society of Agronomy, Madison, pp 434–446

Fischer G, Frohberg K, Parry ML, Rosenzweig C (1996) The potential effects of climate change on world food production and security. In: Bazzaz F, Sombroek W (eds) Global climate change and agricultural production. Wiley, New York, pp 199–235

Fischer G, Shah M, Tubiello FN, Van Velhuizen H (2005) Socio-economic and climate change impacts on agriculture: an integrated assessment, 1990–2080. Philos Trans R Soc Lond B Biol Sci 360:2067–2083

Ford JD, Pearce T, Duerden F, Furgal C, Smit B (2010) Climate change policy responses for Canada's Inuit population: the importance of and opportunities for adaptation. Global Environ Change 20:177–191

Fowler WB, Lopushinsky W (1989) An economical, digital meter for gypsum soil moisture blocks. Soil Sci Am J 53:302–305

Fox P, Rockström J (2003) Supplemental irrigation for dry-spell mitigation of rainfed agriculture in the Sahel. Agric Water Manag 61:29–50

Francia MJR, Durán ZVH, Martínez RA (2006) Environmental impact from mountainous olive orchards under different soil-management systems (SE Spain). Sci Total Environ 358:46–60

Frangi JF, Richard DC, Chavanne X, Bexi I, Sagnard F, Guilbert V (2009) New in situ techniques for the estimation of the dielectric properties and moisture content of soils. Comptes Rendus Geosci 341:831–845

Fuentes S, Collins M, Rogers G, Acevedoe C, Conroy J (2009) Nocturnal heat pulse sap flow as a sensitive system to assess drought effects on grapevines: an irrigation scheduling application? Acta Hortic 646:167–176

Fuhrer J (2003) Agroecosystem responses to combinations of elevated CO2, ozone, and global climate change. Agric Ecosyst Environ 97:1–20

Fujimori S, Matsuoka Y (2007) Development of estimating method of global carbon, nitrogen, and phosphorus flows caused by human activity. Ecol Econ 62:399–418

García J, Romero P, Botía P, García F (2004) Cost-benefits analysis of almond orchard under regultated defict irrigation (RDI) in SE Spain. Spanish J Agric Res 2:157–165

García-Tejero I, Jiménez JA, Reyes MC, Carmona A, Pérez R, Muriel JL (2008) Aplicación de caudales limitados de agua en plantaciones de cítricos del valle del Guadalquivir. Fruticultura Profesional 173:5–16

García-Tejero I, Romero R, Muriel JL (2009) Aplicación de estrategias de riego deficitario y uso de sensores fisiológicos en la gestión del riego en el cultivo de cítricos. Vida Rural 298:26–30

García-Tejero I, Jiménez-Bocanegra JA, Martínez G, Romero R, Durán-Zuazo VH, Muriel-Fernández JL (2010a) Positive impact of regulated deficit irrigation in a commercial citrus orchard (Citrus sinensis L.) Osb. cv. Salustiano. Agric Water Manag 97:614–622

García-Tejero I, Romero R, Jiménez-Bocanegra JA, Martínez G, Durán-Zuazo VH, Muriel JL (2010b) Response of citrus trees to deficit irrigation during different phenological periods in relation to yield, fruit quality, and water productivity. Agric Water Manag 98:689–699

García-Tejero I, Jiménez-Bocanegra JA, Durán-Zuazo VH, Romero-Vicente R, Muriel JL (2010c) Positive impact of deficit irrigation on physiological response and fruit yield in Citrus orchards: implications for sustainable water savings. J Agric Sci Tech 4(3):38–44

Geerts S, Raes D (2009) Deficit irrigation as an on-farm strategy to maximize crop water productivity in dry areas. Agric Water Manag 96:1275–1284

Gelly M, Recasens I, Girona J, Mata M, Arabones A, Rufat J, Marsal J (2003) Effects of water deficit during stage II of peach fruit development and postharvest on fruit quality and ethylene production. J Hort Sci Biotech 78:324–330

Gijón MC, Guerrero J, Couceiro JF, Moriana A (2009) Deficit irrigation without reducing yield or nut splitting in pistachio (*Pistacia vera* cv Kerman on *Pistacia terebinthus* L.). Agric Water Manag 96:12–22

Giménez C, Fereres E, Ruz C, Orgaz F (1997) Water relations and gas exchange of olive trees: diurnal and seasonal patterns of leaf water potential, photosynthesys and stomatal conductance. Acta Hortic 449:411–415

Ginestar C, Castel JR (1996) Responses of young Clementine citrus trees to water stress during different phenological periods. J Hortic Sci 74:551–559

Giraldez C, Fox G (1995) An economic analysis of groundwater contamination from agricultural nitrate emissions in southern Ontario. Can J Agric Econ 43:387–402

Girona J, Marsal J, Mata M, Arbonés A, Mata A (2002) The combined effect of fruit load and water stress in different peach fruit growth stages (*Prunas persica* L.). Acta Hortic 584:149–152

Girona J, Marsal J, Mata M, Arabonés A, DeJong TM (2004) A comparison of the combined effect of water stress and crop load on fruit growth during different phenological stages in young peach trees. J Hort Sci Biotech 79:308–315

Girona J, Mata M, Marsal J (2005) Regulated deficit irrigation during the kernel filling period and optimal irrigation rates in almond. Agric Water Manag 75:152–167

Girona J, Mata M, Del Campo J, Arbones A, Bartra E, Marsal J (2006) The use of midday leaf water potencial for scheduling deficit irrigation in vineyards. Irri Sci 24:115–127

Gisladottir G, Stocking M (2005) Land degradation control and its global environmental benefits. Land Degr Develop 16:99–112

Gleick PH (1993) Water in crisis: a guide to the world's fresh water resources. Oxford University Press, New York, p 474

Gleick PH (2002) Soft water paths. Nature 418:373

Glennon R (2002) The perils of groundwater pumping. Issues Sci Technol 19:73–79

Goldhamer DA, Viveros M (2000) Effects of preharvest irrigation cutoff durations and postharvest water deprivation on almond tree performance. Irrig Sci 19:125–131

Goldhamer DA, Fereres E, Mata M, Girona J, Cohen M (1999) Sensitivity of continuous and discrete plant and soil water status monitoring in peach trees subjected to deficit irrigation. J Am Soc Hortic Sci 124:437–444

Goldhamer DA, Salinas M, Crisosto C, Day KR, Soler M, Moriana A (2002) Effects of regulated irrigation and partial root zone drying on late harvest peach tree performance. Acta Hort 592:343–350

Gómez RA, Salvador MD, Moriana A, Pérez LD, Olmedilla N, Ribas F, Fregapane G (2005) Influence of different irrigation strategies in a cornicabra cv. Olive rochard on virgin olive oil composition and quality. J Food Chem 100:568–575

González AP, Castel JR (2000) Effects of regulated deficit irrigation in 'Clementina de Nules' citrus trees growth, yield and fruit quality. Acta Hort 537:749–758

González AP, Castel JR (2003) Riego deficitario controlado en clementina de nules I. Efectos sobre la producción y la calidad de la fruta, Span. J Agricul Res 1:81–92

González AP, Pavel EW, Oncins JA, Doltra J, Cohen M, Paço T, Massai R, Castel JR (2008) Comparative assessment of five methods of determining sap flow in peach trees. Agric Water Manag 95:503–515

Granier A (1985) Une nouvelle méthode pour la mesure du flux de sève brute dans le tronc des arbres. Annales des Sciences Forestières 42:193–200

Grattan SR, Berenguer MJ, Conell JH, Polito VS, Vossen PM (2006) Olive oil production as influenced by different quantities of applied water. Agric Water Manag 85:133–140

Green SR (1998) Flow by the heat pulse method. HortResearch Internal Report 1998/22. HortResearch, Palmerston North, New Zealand

Green SR, Clothier BE, McLeod DJ (1998) The response of sap flow in apple roots to localised irrigation. Agric Water Manag 33:63–78

Green SR, Clothier B, Jardine B (2003) Theory and practical application of heat pulse to measure sap flow. Agron J 95:1371–1379

Green SR, Clothier B, Peire E (2009) A Re-analysis of heat pulse theory across a wide range of sap flows. In: Fernández E, Diaz-Espejo A (eds) Proceedings of the VIIth IW on sap flow. Acta Hort 846, ISHS 2009

Grewal HS (2010) Water uptake, water use efficiency, plant growth and ionic balance of wheat, barley, canola and chickpea plants on a sodic vertosol with variable subsoil NaCl salinity. Agric Water Manag 97:148–156

Guenette M (2001) Measurements of water erosion and infiltration in Alberta using a rainfall simulator. http://www.agric.gov.ab.ca/sustain/rainfall2.html

Haileslassie A, Peden D, Gebreselassie S, Amede T, Descheemaeker K (2009) Livestock water productivity in mixed crop–livestock farming systems of the Blue Nile basin: assessing variability and prospects for improvement. Agric Syst 102:33–40

Hallikainen MT, Ulaby FT, Dobson MC, el-Rayes MA, Wu LK (1985) Microwave dielectric behaviour of wet- soil. Part I Empirical models and experimental observations. Trans Geosci Remote Sens 26:2311–2316

Hamdy A, Katerji N (2006) Water crisis in the Arab World, analysis and solutions, IAM-Bari Editor, p 60

Hamdy A, Lacirignola C (1999) Mediterranean water resources: major challenges towards the 21st century, IAM Editions. Bari, Italy, p 570

Hamdy A, Sardo V, Farrag GKA (2005) Saline water in supplemental irrigation of wheat and barley under rainfed agriculture. Agric Water Manag 78:122–127

Hamza MA, Anderson WK (2005) Soil compaction in cropping systems: a review of the nature, causes and possible solutions. Soil Till Res 82:121–145

Hantush MM, Mariño MA, Islam MR (2000) Models for leaching of pesticides in soils and groundwater. J Hydrol 227:66–83

Haskett JD, Pachepsky YA, Acock B (2000) Effect of climate and atmospheric change on soybean water stress: a study of Iowa. Ecol Model 135:265–277

Hassanli AM, Ahmadirad S, Beecham S (2010) Evaluation of the influence of irrigation methods and water quality on sugar beet yield and water use efficiency. Agric Water Manag 97:357–362

Heermann DF, Wallender WW, Bos MG (1990) Irrigation efficiency and uniformity. In: Hoffman GS, Howell TA, Soloman KH (eds) Management of farm irrigation system. ASAE, St. Joseph, MI pp 125–149

Hendry K, Sambrook H, Underwood C, Waterfall R, Williams A (2006) Eutrophication of Tamar Lakes (1975–2003): a case study of land-use impacts, potential solutions and fundamental issues for the water framework directive. Water Environ J 20:159–168

Henggeler JC (1998) Managing cotton when water is limited. In: Proceedings of the beltwide cotton conference, National Cotton Council, Memphis, Tennessee, pp 641–645

Herweg K, Ludi E (1999) The performance of selected soil and water conservation measures—case studies from Ethiopia and Eritrea. Catena 36:99–114

Herweg K, Steiner K (2002a) Impact monitoring and assessment. Instruments for use in rural development projects with a focus on sustainable land management. In: Procedure, Centre for Development and Environment (CDE, Switzerland). Deutsche Gesellschaft für Technische Zusammenarbeit (GTZ, Germany), Swiss Agency for Development and Cooperation (SDC, Switzerland), Intercooperation (Switzerland), Helvetas (Switzerland), Rural Development Department of the World Bank, Wabern, Switzerland, vol 1

Herweg K, Steiner K (2002b) Impact monitoring and assessment. Instruments for use in rural development projects with a focus on sustainable land management. In: Toolbox, Centre for Development and Environment (CDE, Switzerland), Deutsche Gesellschaft für Technische Zusammenarbeit (GTZ, Germany), Swiss Agency for Development and Cooperation (SDC, Switzerland), Intercooperation (Switzerland), Helvetas (Switzerland), Rural Development Department of the World Bank, Wabern, Switzerland, Vol 2

Higgs KH, Jones HG (1984) A microcomputer-based system for continuous measurement and recording fruit diameter in relation to environmental factors. J Exp Bot 35:1646–1655

Hillel D (1991) Out of the earth: civilization and the life of the soil. The Free Press, New York

Hillel D (2008) Soil in the environment. Cruciable of terrestrial life. Academic Press, Amsterdam, p 307

Hinrichsen D, Robey B, Upadhyay UD (1998) Solutions for a water-short world. Population Reports series M. no. 14. Johns Hopkins School of Public Health, Population Information Program, Baltimore

Hodges AW, Lynne GD, Rahmani M, Casey CF (1994) Adoption of energy and water-conserving irrigation technologies in Florida, Fact Sheet EES 103, Florida Cooperative Extension Service, Institute of Food and Agricultural Sciences, University of Florida, USA

Hoff H (2009) Global water resources and their management. Curr Opin Environ Sustainability 1:141–147

Howarth RW, Billen G, Swaney D, Townsend A, Jaworski N, Lajtha K, Downing JA, Elmgren R, Caraco N, Jordan T, Berendse F, Freney J, Kudeyarov V, Murdoch P, Zhao-Liang Z (1996) Regional nitrogen budgets and riverine N and P fluxes for the drainages to the North Atlantic Ocean: natural and human influences. Biogeochemistry 35:75–139

Howell TA (2001) Enhancing water use efficiency in irrigated agriculture. Agron J 93:281–289

Howell TA, Schneider AD, Jensen ME (1991) History of lysimeter design and use for evapotranspiration measurements. In: Allen RG (ed) Lysimeters for evapotranspiration and environmental measurements. Am Soc Civ Eng, Reston, pp 1–9

Hsiao TC (1973) Plant response to water stress. Ann Rev Plant Physiol 24:519–570

Hsiao TC (1993) Effects of drought and elevated CO2 on plant water use efficiency and productivity. In: Jackson MB, Black CR (eds) Global environment change: interacting stresses on plants in a changing climate NATO ASI, vol Series I. Springer, Berlin, pp 435–465

Huang J, Hu R, Van Meijl H, Van Tongeren F (2004) Biotechnology boosts to crop productivity in China: trade and welfare implications. J Develop Econ 75:27–54

Huang Q, Rozelle S, Lohmar B, Huang J, Wang J (2006) Irrigation, agricultural performance and poverty reduction in China. Food Policy 31:30–52

Hubbard B, Gelting R, Baffigo V, Sarisky J (2005) Community environmental health assessment strengthens environmental public health services in the Peruvian Amazon. Int J Hyg Environ Health 208:101–107

Hueso JJ, Cuevas J (2010) Ten consecutive years of regulated deficit irrigation probe the sustainability and profitability of this water saving strategy in loquat. Agric Water Manag 97:645–650

Hussain I, Mudasser M, Hanjra MA, Amrasinghe U, Molden D (2004) Improving wheat productivity in Pakistan: econometric analysis using panel data from Chaj in the upper Indus Basin. Water Int 29:189–200

Hutmacher RB, Nightingale HI, Rolston DE, Biggar JW, Dale F, Vail SS, Peters D (1994) Growth and yield responses of almond *(Prunus-Amygdalus)* to trikle irrigation. Irrig Sci 14:117–126

Hutmacher RB, Phene CJ, Davis KR, Vail SS, Kerby TA, Peters M, Hawk CA, Keeley M, Clark DA, Ballard D, Hudson N (1995) Evapotranspiration, fertility management for subsurface drip Acala and Pima cotton. In: Lamm FR (ed) Proceedings of the 5th International Microirrigation Congress. St. Joseph, Michigan, pp 147–154

Huygen J, Van den Broek BJ, Rabat P (1995) Hydra Model Trigger, a soil éter balance and crop growth simulation system for irrigation water management purposes. Paper submitted to ICID/FAO Workshop, September 1995. Rome. Irrigation scheduling: from theory to practise. FAO, Rome, Italy

ICAR (1999) Fifty years of natural resource management research. In: Singh GB, Shama BR (eds) Division of natural resource management. Indian Council of Agricultural Research, New Delhi

IFEN (1997) (French Institute of the Environment). Agriculture and the Environmental Indicators, 1997–1998 (edn). www.ifen.fr

Igbadun HE, Tarimo AKPR, Salim BA, Mahoo HF (2007) Evaluation of selected crop water production functions for an irrigated maize crop. Agric Water Manag 94:1–10

Imtiyaz M, Mgadla NP, Chepete B, Manase SK (2000) Response of six vegetable crops to irrigation schedules. Agric Water Manag 45:331–342

Intrigliolo DS, Chirivella C, Castel, JR (2005) Response of grapevine cv. Tempranillo to irrigation amount and partial rootzone drying under contrasting crop load levels. In: International symposium on advances in grapevine and wine research. Congress ISHS Venosa, Italy

IPCC (2001a) Intergovernmental Panel on Climate Change. Climate Change 2001: The Scientific Basis, Cambridge University Press, Cambridge, UK, 881 pp

IPCC (2001b) Intergovernmental Panel on Climate Change, Climate Change 2001: Impacts, Adaptation, and Vulnerability, Cambridge University Press, Cambridge, UK, 1032 pp

Iqbal MM, Shah SM, Mohammad W, Nawaz H (1999) Field response of potato subjected to water stress at different growth stages. In: Kirda C, Moutonnet P, Hera C, Nielsen DR (eds) Crop yield response to deficit irrigation. Kluwer Academic Publishers, Dordrecht, pp 213–223

Irmak S, Haman DZ, Bastug R (2000) Determination of crop water stress index for irrigation timing and yield estimation of corn. Agron J 92:1221–1227

Itier B, Maraux F, Ruelle P, Deumier JM (2010) Applicability and limitations of irrigation scheduling methods and techniques. FAO Corporate Document Repository. http://www.fao.org/docrep/w4367e/w4367e04.htm

IWMI (2001) International Water Management Institute Sustainable Groundwater Management Theme, International Water Management Institute, CGIAR. Consultative Group on International Agricultural Research

Jackson RB, Carpenter SR, Dahm CN, McKnight DM, Naiman RJ, Postel SL, Running SW (2002) Water in a changing world. Ecol Appl 11:1027–1045

Jalota SK, Sood A, Chahal ABS, Choudhury BU (2006) Crop water productivity of cotton (*Gossypum hirsutum* L.)-wheat (*Triticum aestivum* L.) system as influenced by deficit irrigation, soil texture and precipitation. Agric Water Manag 84:137–146

Jara J, Stockle CO, Kjelgaard J (1998) Measurement of evapotranspiration and its components in a corn (*Zea Mays* L.) field. Agric Forest Meteorol 92:131–145

Johnstone PR, Hartz TK, LeStrange M, Nunez JJ, Miyao EM (2005) Managing fruit soluble solids with late-season deficit irrigation in drip-irrigated processing tomato production. HortScience 40:1857–1861

Jones HG (2004) Irrigation scheduling: advantages and pitfalls of plant-based methods. J Exp Bot 407:2427–2436

Jones PG, Thornton PK (2003) The potential impacts of climate change on maize production in Africa and Latin American in 2055. Global Environ Change 13:51–59

Kabore D, Reij C (2004) The emergence and spreading of an improved traditional soil and water conservation practice in Burkina Faso. International Food Policy Research Institute, Washington DC. http://www.ifpri.org/divs/eptd/dp/papers/

Kahinda JMM, Rockström J, Taigbenu AE, Dimes J (2007) Rainwater harvesting to enhance water productivity of rainfed agriculture in the semi-arid Zimbabwe. Phys Chem Earth Parts 32:1068–1073

Kang S, Zhang J (2004) Controlled alternate partial root-zone irrigation: its physiological consequences and impact on water use efficiency. J Exp Bot 55:2437–2446

Kang S, Shi W, Zhang J (2000) An improved water-use efficiency for maize grown under regulated deficit irrigation. Field Crops Res 67:207–214

Kang S, Hu X, Goodwin I, Jerie P (2002) Soil water distribution, water use and yield response to partial rootzone drying under a shallow groundwater table condition in a pear orchard. Sci Hortic 92:277–291

Kashyap PS, Panda PK (2003) Effect of irrigation scheduling on potato crop parameters under water stressed conditions. Agric Water Manag 59:49–66

Katerji N, Bethenod O (1997) Comparaison du comportement hydrique et de la capacit'e photosynth'etique du mais et du tournesol en condition de contrainte hydrique. Conclusions sur l'efficience de l'eau. Agronomie 17:17–24

Katerji N, Perrier A (1985) Determination de la resistance globale d'un couvert vegetal a la diffusion de vapeur d'eau et de ses differentes composantes. Approche théorique et vérification expérimentale sur une culture de luzerne. Agric Meteorol 34:105–120

Katerji N, Mastrorilli M, Rana G (2008) Water use efficiency of crops cultivated in the Mediterranean region: Review and análisis. Eur J Agron 28:493–507

Kayombo B, Hatibu N, Mahoo HF (2004) Effect of micro-catchment in rainwater harvesting on yield of maize in a semi-arid area. In: 13th international soil conservation organisation conference. Conserving soil and water for society, sharing solutions, Brisbane. http://www.tucson.ars.ag.gov/isco/isco13/PAPERS%20F-L/KAYOMBO.pdf

Keller M (2005) Deficit irrigation and vine mineral nutrition. Am J Enol Vitic 56:267–283

Kennedy AC, Gewin VL (1997) Soil microbial diversity: present and future considerations. Soil Sci 162:607–617

Khan S, Hanjra MA (2008) Sustainable land and water management policies and practices: a pathway to environmental sustainability in large irrigation systems. Land Degrad Develop 19:469–487

Khan S, Tariq R, Yuanlai C, Blackwell J (2006) Can irrigation be sustainable? Agric Water Manag 80:87–99

Kirda C, Cetin M, Dasgan Y, Topcu S, Kaman H, Ekici B, Derici MR, Ozguven AI (2004) Yield response of greenhouse grown tomato to partial root drying and conventional deficit irrigation. Agric Water Manag 69:191–201

Klocke G, Kranz WL, Hubbard K, Watts DG (1996) Irrigation. Cooperative Extension, Institute of Agriculture and Natural Resources, Lincoln, NE, University of Nebraska. http://www.ianr.unl.edu/pubs/irrigation/g992.htm

Kodad O, Socias R (2006) Influence of genotype, year and type of fruiting branches on the productive behaviour of almond. Sci Hortic 109:297–302

Koulouri M, Giourga Ch (2007) Land abandonment and slope gradient as key factors of soil erosion in Mediterranean terraced lands. Catena 69:274–281

Koundouras S, Marinos V, Gkoulioti A, Kotseridis Y, Van Leeuwen C (2006) Influence of vineyard location and vine water status on fruit maturation of nonirrigated cv Agiorgitiko (Vitis vinifera L.). Effects on wine phenolic and aroma components. J Agric Food Chem 54:5077–5086

Krüger E, Schmidt G, Brückner U (1999) Scheduling strawberry irrigation based upon tensiometer measurement and a climatic water balance model. Sci Hortic 81:409–424

Kunze D (2000) Methods to evaluate the economic impact of water harvesting. Q J Int Agric 39:69–91

La Gal LS, Marlin C, Leduc C, Taupin C, Massult JD, Favreau G (2001) Renewal rate estimation of groundwater based on radioactive tracers (^{3}H, ^{14}C) in an unconfirmed aquifer in a semi-arid area, Iullemeden Basin. Niger J Hydrol 254:145–156

Laboski CAM, Lamb JA, Dowdy RH, Baker JM, Wright J (2001) Irrigation scheduling for a sandy soil using mobile frequency domain reflectometry with a checkbook method. J Soil Water Conserv 56:97–100

Lagacherie P, Coulouma G, Ariagno P, Virat P, Boizard H, Richard G (2006) Spatial variability of soil compaction over a vineyard region in relation with soils and cultivation operations. Geoderma 134:207–216

Lal R (1993) Soil erosion and conservation. In: Pimentel D (ed) World soil erosion and conservation. Cambridge University Press, Cambridge, pp 7–26

Lal R (2001) Potential of desertification control to sequester carbon and mitigate the greenhouse effect. Clim Change 51:35–72

Lal R (2004) Carbon sequestration in dryland ecosystems. Environ Manag 33:528–544

Lal R (2007) anthropogenic influences on world soils and implications to global food security. Adv Agron 93:69–93

Lawlor DW, Mitchell RAC (2000) Crop ecosystem responses to climatic change: wheat. In: Reddy KR, Hodges HF (eds) Climate change and global crop productivity. CABI Pub, New York, pp 57–80

Lea CJD, Syvertsen JP (1993) Salinity reduces water use and nitrate-n-use efficiency of citrus. Ann Bot 72:47–54

Ledieu J, De Ridder O, De Clerck P, Dautrebande S (1986) A method of measuring soil moisture by time-domain reflectometry. J Hydrol 88:319–328

Leib BG, Caspari HW, Redulla CA, Andrews PK, Jabro JJ (2006) Partial root zone drying and deficit irrigation of 'Fuji' apples in a semi-arid climate. Irrig Sci 24:541–552

Lemaire G, Van Oosterom E, Sheehy J, Jeuffroy MH, Massignam A, Rossato L (2007) Is crop N demand more closely related to dry matter accumulation or leaf area expansion during vegetative growth? Field Crops Res 100:91–106

Leuning R (1995) A critical appraisal of a combined stomata photosynthesis model for C3 plants. Plant Cell Environ 18:339–355

Li XY (2003) Rainwater harvesting for agricultural production in the semiarid loess region of China. Food Agric Environ 1:282–285

Li XY, Gong JD (2002) Compacted microcatchments with local earth materials for rainwater harvesting in the semiarid region of China. J Hydrol 257:134–144

Li SH, Huguet JG, Schoch PG, Orlando P (1989) Response of peach tree growth and cropping to soil water deficit at various phenological stages of fruit development. J Hort Sci 64:541–552

Li F, Zhao S, Geballe GT (2000) Water use patterns and agronomic performance for some cropping systems with and without fallow crops in a semi-arid environment of northwest China. Agric Ecosyst Environ 79:129–142

Li J, Inanaga S, Li Z, Eneji AE (2005) Optimizing irrigation scheduling for winter wheat in the North China Plain, Agric Water Manag 76:8–23

Li XY, Su D, Yuan Q (2007) Ridge-furrow planting of alfalfa (Medicago sativa L.) for improved rainwater harvest in rainfed semiarid areas in Northwest China. Soil Till Res 93:117–125

Liang J, Zhang J, Wong MH (1997) Can stomatal closure caused by xylem ABA explain the inhibition of leaf photosythesis under soil drying? Photosynth Res 51:149–159

Lioubimtseva E, Cole R, Adams JM, Kapustin G (2005) Impacts of climate and land-cover changes in arid lands of Central Asia. J Arid Environ 62:285–308

Liu GD, Wu WL, Zhang J (2005) Regional differentiation of non-point source pollution of agriculture-derived nitrate nitrogen in groundwater in northern China. Agric Ecosyst Environ 107:211–220

Liu F, Shahnazari A, Andersen MN, Jacobsen SE, Jensen CR (2006) Effects of deficit irrigation (DI) and partial root drying (PRD) on gas exchange, biomass partitioning and water use efficiency in potato. Sci Hortic 109:113–119

Lobell D, Asner G (2003) Climate and management contributions to recent trends in agricultural yields. Science 299:1032

Lord EI, Mitchell R (1998) Effect of nitrogen inputs to cereals on nitrate leaching from sandy soils. Soil Use Manag 14:78–83

Ludwig JA, Wilcox BP, Breshears DD, Tongway DJ, Imeson AC (2005) Vegetation patches and runoff-erosion as interacting ecohydrological processes in semiarid landscapes. Ecology 86:288–297

Lyson TA (2002) Advanced agricultural biotechnologies and sustainable agriculture. Trends Biotechnol 20:193–196

MAF (1997) Indicators of sustainable irrigated agriculture. Ministry of Agriculture and Forestry, MAF Technical Paper No 00/03, Wellington, New Zealand, pp 1–50

Malano HM, Turral HN, Wood ML (1996) Surface irrigation management in real time in southeastern Australia: irrigation scheduling and field application. Irrigation scheduling: from theory to practice. In: Proceedings ICID/FAO workshop, Water Report N° 8. FAO, Rome, Italy, Sept 1995

Maltby E (1991) Wetland management goals: wise use and conservation. Land Urban Plan 20:9–18

Mannion AM (1995a) Agriculture, environment and biotechnology. Agric Ecosyst Environ 53:31–45

Mannion AM (1995b) Agriculture and environmental change, Temporal and spatial dimensions. Wiley, Chichester, p 405

Mansouri-Far C, Modarres-Sanavy SAM, Farhad-Saberaly S (2010) Maize yield response to deficit irrigation during low-sensitive growth stages and nitrogen rate under semi-arid climatic conditions. Agric Water Manag 97:12–22

Margat J, Vallée D (1997) Démographie en Méditerranée. Options Méditerranéennes Série A 31:3–16

Mariolakos I (2007) Water resources management in the framework of sustainable development. Desalination 213:147–151

Marouelli WA, Silva WLC (2007) Water tension thresholds for processing tomatoes under drip irrigation in central Brazil. Irrig Sci 25:411–418

Marouelli WA, Silva WLC, Moretti CL (2004) Production, quality and water use efficiency of processing tomato as affected by the final irrigation timing. Hort Brasil 22:225–230

Marsal J, Rapoport HF, Manrique T, Girona J (2000) Pear fruit growth under regulated deficit irrigation in container-growth plants. Sci Hortic 85:243–259

Marsal J, Mata M, Arabonés A, Rufat J, Girona J (2002) Regulated deficit irrigation and rectification of irrigation scheduling in young pear trees: An evaluation based on vegetative and productive response. Eur J Agron 17:111–122

Matson PA, Parton WJ, Power AG, Swift MJ (1997) Agricultural intensification and ecosystem properties. Science 277:504–509

Matsumoto K, Ohta M, Tanaka T (2005) Dependence of stomatal conductance on leaf chlorophyll concentration and meteorological variables. Agric Forest Meteorol 132:44–57

McCarthy JJ, Canziani OF, Leary NA, Dokken DJ, White KS (2001) Climate Change 2001 impacts, adaptation and vulnerability. Contribution of working group II to the third assessment report of the intergovernmental panel on climate change. Cambridge University Press, Cambridge

McCarthy MG, Loveys BR, Dry PR, Stoll M (2002) Regulated deficit irrigation and partial rootzone drying as irrigation management techniques for grapevines. In: Deficit irrigation practices, FAO Water Reports No. 22, Rome Italy, FAO, pp 79–87

McCray K (2001) American agriculture and ground water: brining issues to the surface. Irrig J 51:27–29

Medrano H, Escalona JM, Cifre J, Bota J, Flexas J (2003) A ten-year study on the physiology of two Spanish grapevine cultivars under field conditions: effects of water availability from leaf photosynthesis to grape yield and quality. Funct Plant Biol 30:607–619

Meerkerk A, Van Wesemael B, Cammeraat E (2008) Water availability in almond orchards on marl soils in southeast Spain: the role of evaporation and runoff. J Arid Environ 72:2168–2178

Melloul AJ, Collin ML (2003) Harmonizing water management and social needs: a necessary condition for sustainable development. The case of Israel's coastal aquifer. J Environ Manag 67:385–394

Merot A, Wery J, Isbérie C, Charron F (2008) Response of a plurispecific permanent grassland to border irrigation regulated by tensiometers. Eur J Agron 28:8–18

Metzidakis I, Martinez VA, Castro NG, Basso B (2008) Intensive olive orchards on sloping land: good water and pest management are essential. J Environ Manag 89:120–128

Michelakis N, Vouyoukalou E, Clapaki G (1996) Water use and soil moisture depletion by olive trees under different irrigation conditions. Agric Water Manag 29:315–325

Mitchell PD, Chalmers DJ, Jerie PH, Burge G (1986) The use of initial withholding of irrigation and tree spacing to enhance the effect of regulated deficit irrigation on pear trees. J Am Soc Hort Sci 111:858–861

Mitchell PD, Van den Ande B, Jerie PH, Chalmers DJ (1989) Response of 'Barlett' pear to withholding irrigation, regulated deficit irrigation, and tree spacingx. J Am Soc Hort Sci 114:15–19

Mohamed YY (2002) Water harvesting and spreading in the arid zones of Sudan. In: Dupuy A, Lee C, Schaaf T (eds) In: Proceedings of the international seminar on combating desertification: freshwater resources and the rehabilitation of degraded areas in the drylands. Samantha Wauchope, Fushia publishing Paris, N'Djamena, Chad, pp 27–28

Mohammed MJ, Zuraiqi S, Quasmch Q, Papadoulos R (1999) Yield response and nitrogen utilization efficiency by drip irrigated potato. Nutr Cycl Agroecosyst 54:243–249

Molden D, Oweis T, Steduto P, Kijne JW, Hanjra MA, Bindraban PS (2007) Pathways for increasing agricultural water productivity, In: Water for food, water for life: a comprehensive assessment of water management in agriculture, International Water Management Institute, London, Earthscan, Colombo, Sri Lanka

Molenat J, Gascuel OC (2002) Modelling flow and nitrate transport in groundwater for the prediction of water travel times and of consequences of land use evolution on water quality. Hydrol Proc 16:479–492

Monirul QMM (2003) Climate change and extreme weather events: can developing countries adapt? Climate Policy 3:233–248

Moreno F, Cayuela JA, Fernández JE, Fernández BE, Murillo JM, Cabrera F (1996a) Water balance and nitrate leaching in an irrigated maize crop in SW Spain. Agric Water Manag 32:71–83

Moreno F, Fernández JE, Clothier BE, Green SR (1996b) Water use and root uptake by olive trees. Plant Soil 184:84–96

Moreno F, Conejero W, Martín PMJ, Girón IF, Torrecillas A (2006) Maximum daily trunk shrinkage referente values for irrigation scheduling in olive trees. Agric Water Manag 84:290–294

Moriana A, Fereres E (2002) Plant indicators for scheduling irrigation of young olive trees. Irrig Sci 21:83–90

Moriana A, Orgaz F, Pastor M, Fereres E (2003) Yield responses of mature olive orchard to water deficits. J Am Soc Hort Sci 123:425–431

Mpelasoka BS, Behboudian MH (2002) Production of aroma volatiles in response to deficit irrigation an to crop load in relation to fruit maturity for 'Braeburn' apple. Postharvest Biol Tec 24:1–11

Mpelasoka BS, Behboudian MH, Dixon J, Neal SM, Caspari HW (2000) Improvement of fruit quality and storage potential of 'Braeburn' apple through deficit irrigation. J Hort Sci Biotech 75:615–621

Mpelasoka BS, Behboudian MH, Green SR (2001) Water use, yield and fruit quality of lysimeter-grown apple trees: responses to deficit irrigation and to crop load. Irrig Sci 20:107–113

Müller K, Magesan GN, Bolan NS (2007) A critical review of the influence of effluent irrigation on the fate of pesticides in soil. Agric Ecosyst Environ 120:93–116

Muñoz CM (2005) El cultivo del olivo con riego localizado. Diseño y manejo del cultivo y las instalaciones. Programación de riegos y fertirrigación. Mundi-Prensa, Madrid

Murgai R, Ali M, Byerlee D (2001) Productivity growth and sustainability in post green-revolution agriculture: the case of Indian and Pakistani Punjabs. World Bank Research Observer 16:199–218

Muriel JL, García-Tejero I, Romero R, Ruiz R, Cristino L, Pinto L, Hernández A (2009) Physiological response of citrus trees to different strategies of deficit irrigation. Books of Abstracts ICHS, Viña del Mar, Chile, p 80

Myers N, Kent J (2001) Perverse subsidies: how tax dollars can undercut the environment and the economy. Island Press, Washington DC

Nadezhdina N (1999) Sap flow index as an indicator of plant water status. Tree Physiol 19:885–891

Nadezhdina N, Čermák J (1997) Automatic control unit for irrigation systems based on sensing the plant water status. An Inst Sup Agron 46:149–157

Nanos GD, Kazantzis L, Kefalas P, Petrakis C, Stavroulakis GG (2002) Irrigation and harvest time affect almond kernel quality and composition. Sci Hortic 96:249–256

Naor A (2000) Midday stem water potential as a plant water stress indicator for irrigation scheduling in fruit trees. Acta Hortic 537:447–454

Naor A (2006) Irrigation scheduling and evaluation of tree water status in deciduous orchards. Hort Rev 32:111–165

Naor A, Cohen S (2003) Sensitivity and variability of maximum trunk shrinkage, midday stem water potential, and transpiration rate in response to withholding irrigation from field grown apple trees. HortScience 38:547–551

Naor A, Hupert H, Greenblat Y, Peres M, Klein I (2001) The response of nectarine fruit size and midday stem water potential to irrigation level in stage III and crop load. J Am Soc Hort Sci 126:140–143

Naor A, Peres M, Greenblat Y, Gal Y, Ben-Arie R (2004) Effects of pre-harvest irrigation regime and crop level on yield, fruit size distribution and fruit quality of field-grown 'Black Amber' Japanese plum. J Hort Sci Biotech 79:281–288

Ngigi SN (2003a) Rainwater harvesting for improved food security: promising technologies in the greater Horn of Africa. GHARP, KRA, Nairobi

Ngigi SN (2003b) What is the limit of up-scaling rainwater harvesting in a river basin? Phys Chem Earth 28:943–956

Nicolas E, Torrecillas A, Ortuño MF, Domingo R, Alarcón JJ (2005) Evaluation of transpiration in adult apricot trees from sap flow measurements. Agric Water Manag 72:131–145

Nixon SW (1995) Managing wastewater in coastal urban areas. Ophelia 41:196–198

Nji A, Fonteh MF (2002) Potential and constraints to the adoption of water harvesting technologies in traditional economies. Discov Innov 14:202–214

Noborio K (2001) Measurement of soil water content and electrical conductivity by time domain reflectometry: a review. Comput Electron Agric 31:213–237

Nortes PA, Pérez PA, Egea G, Conejero W, Domingo R (2005) Comparison of changes in item diameter and water potential values for detecting water stress in young almond trees. Agric Water Manag 77:296–307

Novák V, Hurtalová T, Matejka F (2005) Predicting the effects of soil water content and soil water potential on transpiration of maize. Agric Water Manag 76:211–223

Odum TH (1983) Systems ecology: an introduction. Wiley, New York

Ojeda H, Andary C, Kraeva E, Carbonneau A, Deloire A (2002) Influence of preand postveraison water deficit on synthesis and concentration of skin phenolic compounds during berry growth of Vitis vinifera cv. Shiraz, Am J Enol Vitic 53:261–267

Ojeda IM, Mayer SA, Solomon DB (2008) Economic valuation of environmental services sustained by water flows in the Yaqui River Delta. Ecol Econo 65:155–166

Oki T, Agata Y, Kanae S, Saruhashi T, Yang D, Musiake K (2001) Global assessment of current water resources using total runoff integrating pathways. Hydrol Sci J 46:983–995

Olesen JE, Bindi M (2002) Consequences of climate change for European agricultural productivity, land use and policy. Eur J Agron 16:239–262

Orgaz F, Fereres E (2004) Riego. In: Barranco D, Fernández-Escobar R, Rallo L (eds) El cultivo del olivo. Mundi-Prensa, Madrid, pp 425–431

Ortuño MF, Alarcón JJ, Nicolás E, Torrecillas A (2004) Interpreting trunk diameter changes in young lemon trees under deficit irrigation. Plant Sci 167:275–280

Ortuño MF, Conejero W, Moreno F, Moriana A, Intrigliolo DS, Biel C, Mellisho CD, Pérez PA, Domingo R, Ruiz SMC, Casadesus J, Bonany J, Torrecillas A (2010) Could trunk diameter sensors be used in woody crops for irrigation scheduling? A review of current knowledge and future perspectives. Agric Water Manag 97:1–11

Oster JD, Wichelns D (2003) Economic and agronomic strategies to achieve sustainable irrigation. Irrig Sci 22:107–120

Ouedraogo M, Kaboré V (1996) The "zai": a traditional technique for the rehabilitation of degraded land in the Yatenga, Burkina Faso. In: Reij C, Scoones I, Toulmin C (eds), Sustaining the soil. Indigenous soil and water conservation in Africa, pp 80–92

Oweis T (1997) Supplemental irrigation. A highly efficient water use practice, ICARDA editions, p 16

Oweis T, Hachum A (2001) Reducing peak supplemental irrigation demands by extending sowing dates. Agric Water Manag 10:21–34

Oweis T, Hachum A, Kijne J (1999) Water harvesting and supplemental irrigation for improved water use efficiency in dry areas. International Water Management Institute, Colombo, pp 1–51

Oweis T, Hachum A, Pala M (2005) Faba bean productivity under rainfed and supplemental irrigation in northern Syria. Agric Water Manag 73:57–72

Paavola JJ (2008) Livelihoods, vulnerability and adaptation to climate change in Morogoro, Tanzania. Environ Sci Policy 11:642–654

Pala M, Ryan J, Zhang H, Singh M, Harris HC (2007) Water-use efficiency of wheat-based rotation systems in a Mediterranean environment. Agric Water Manag 93:136–144

Palmer PL (2001) The Pacific Northwest cooperative agricultural weather network. U.S. Bureau of Reclamation, Pacific Northwest Region. 1150 N. Curtis Road, Suite 100, Boise, Idaho

Pandey DN (2001) A bountiful harvest of rainwater. Science 293:1763

Pandey DN, Gupta AK, Anderson DM (2003) Rainwater harvesting as an adaptation to climate change. Curr Sci 85:46–59

Patakas A, Noitsakis B, Chouzouri A (2005) Optimization of irrigation water use in grapevines using the relationship between transpiration and plant water status. Agric Ecosyst Environ 106:253–259

Patanè C, Cosentino SL (2010) Effects of soil water deficit on yield and quality of processing tomato under Mediterranean climate. Agric Water Manag 97:131–138

Patumi M, d'Andrina R, Fontanazza G, Morelli G, Giorgio P, Sorrentino G (1999) Yield and oil quality of intensively trained trees of three cultivars of olive (olea europea L.) under different irrigation regimes. J Hort Sci Biotech 74:729–737

Payero JO, Melvin SR, Irmak S, Tarkalson D (2006) Yield response of corn to deficit irrigation in a semiarid climate. Agric Water Manag 84:101–112

Pearcy RW (1983) Physiological consequences of cellular water deficits: efficient water use in crop production. In: Taylor HM, Jordan WM, Sinclair TR (eds) Limitations of efficient water use in crop production, Am Soc Agron Inc, Crop Sci Soc Amer Inc, Soil Sci Soc Am Inc, Madison, USA, pp 277–287

Pellegrino A, Lebon E, Simonneau T, Wery J (2005) Towards a simple indicator of water stress in grapevine (Vitis vinifera L.) based on the differential sensitivities of vegetative growth components. Aust J Grape Wine Res 11:306–315

Pereira SL (1999) Higher performance through combined improvements in irrigation methods and scheduling: a discussion. Agric Water Manag 40:153–169

Pimentel D, Kounang N (1998) Ecology of soil erosion in ecosystems. Ecosystem 1:416–426

Pimentel D, Pimentel M (2003) World population, food, natural resources, and survival. World Futures 59:145–167

Pimentel D, Harvey C, Resosudarmo P, Sinclair K, Kurz D, McNair M, Crist S, Shpritz L, Fitton L, Saffouri R, Blair R (1995) Environmental and economic costs of soil erosion and conservation benefits. Science 267:1117–1123

Pimentel D, Herz M, Whitecraft M, Zimmerman M, Allen R, Becker K, Evans J, Hussan B, Sarsfield R, Grosfeld A, Seidel T (2002a) Renewable energy: current and potential issues. BioScience 52:1111–1120

Pimentel D, Doughty R, Carothers C, Lamberson S, Bora N, Lee K (2002b) Energy inputs in crop production: comparison of developed and developing countries. In: Lal L, Hansen D, Uphoff N, Slack S (eds) Food security environmental quality in the developing world. CRC Press, Boca Raton, pp 129–151

Pimentel D, Berger B, Filiberto D, Newton M, Wolfe B, Karabinakis B, Clark S, Poon E, Abbett E, Nandagopal S (2004) Water resources: agriculture, and the environment Environmental Issues. Bioscience 54:909–918

Pingali PL, Marquez CB, Palis FG (1994) Pesticides and Philippine rice farmer health: a medical and economic analysis. Am J Agric Econ 76:587–592

Planton S, Déqué M, Chauvin F, Terray L (2008) Expected impacts of climate change on extreme climate events. Comptes Rendus Geosciences 340:564–574

Playán E, Mateos L (2006) Modernization and optimization of irrigation systems to increase water productivity. Agric Water Manag 80:100–116

Postel S (1999) Pillar of sand. Can the irrigation miracle last?. WW Norton & Company, New York

Pretorius E, Woyessa YE, Slabbert SW, Tetsoane S (2005) Impact of rainwater harvesting on catchment hydrology: a case study of the Modder River basin, South Africa. Trans Ecol Environ 80:1–59

Prinz D, Malik AH (2002) Runoff farming. Article prepared for WCA infoNet. http://www.wca-infonet.org

Qin Z, Su G, Zhang J, Ouyang Y, Yu Q, Li J (2010) Identification of important factors for water vapor flux and CO_2 exchange in a cropland. Ecol Modell 221:575–581

Ragan GE, Young RA, Makela CJ (2000) New evidence on the economic benefits of controlling salinity in domestic water supplies. Water Res 36:1087

Räisänen J (2001) CO2-induced climate change in CMIP2 experiments. Quantification of agreement and role of internal variability. J Climate 14:2088–2104

Ramankutty N, Foley JA, Olejniczak NJ (2002) People and land: changes in global population and croplands during the 20th century. Ambio 31:251–2257

Ramos DE, Weinbaum SA, Shackel KA, Schwankle LJ, Mitcham EJ, Mitchell FG, Synder RG, Mayer G, McGourty G (1994) Influence of tree water status and canopy position on fruit size and quality of 'Barlet' pears. Acta Hortic 367:192–200

Rapp A (1987) Desertification. In: Gregory KJ, Walling DE (eds) Human activity and environmental processes. Wiley, UK, pp 425–443

Raun WR, Johnson GV (1999) Improving nitrogen use efficiency for cereal production. Agron J 91:357–363

Reid MA, Brooks JJ (2000) Detecting effects of environmental water allocations in wetlands of the Murray-Darling Basin, Australia, Regulated Rivers. Res Manag 16:479

Reilly J, Graham J, Hrubovcak J (2001) Agriculture: the potential consequences of climate variability and change for the United States, US National assessment of the potential consequences of climate variability and change, US Global Change Research Program. Cambridge University Press, New York. http://www.usgcrp.gov/usgcrp/Library/nationalassessment/Agriculture.pdf

Richter GM, Semenov MA (2005) Modelling impacts of climate change on wheat yields in England and Wales: assessing drought risks. Agric Syst 84:77–97

Rijsberman F, De Silva S (2006) Sustainable agriculture and wetlands. In: Verhoeven JTA, Beltman B, Bobbink R, Whigham DF (eds) Wetlands and natural resource management. Springer, Berlin, pp 33–52

Ringersma J, Batjes W, Dent DC (2003) Green water definition and data for assessment. The Netherlands, ISRIC

Ritchie JT, Basso B (2008) Water use efficiency is not constant when crop water supply is adequate or fixed: The role of agronomic management. Eur J Agron 28:273–281

Rockstrom J (2003) Water for food and nature in drought-prone Tropics: vapour shift in rain-fed agriculture. Royal Soc Trans Biol Sci 358:1997–2009

Rockström J (1999) On farm green water estimates as a tool for increased food production in water scarce regions. Phys Chem Earth 8:375–383

Rockström J (2004) Making the best of climatic variability: options for upgrading rainfed farming in water scarce regions. Water Sci Technol 49:151–156

Rockström J, Barron J (2007) Water productivity in rainfed systems: overview of challenges and analysis of opportunities in water scarcity prone savannahs. Irrig Sci 25:299–311

Rockström J, Barron J, Fox P (2002) Rainwater management for increased productivity among small-holder farmers in drought prone environments. Phys Chem Earth 27:949–959

Rockström J, Folke C, Gordon L, Hatibu N, Jewitt G, Penning de Vries F, Rwehumbiza F, Sally H, Savenije H, Schulze R (2004) A watershed approach to upgrade rainfed agriculture in water scarce regions through water system innovations: an integrated research initiative on water for food and rural livelihoods in balance with ecosystem functions. Phys Chem Earth 29:1109–1118

Rockström J, Karlberg L, Wani SP, Barron J, Hatibu N, Oweis T, Bruggeman A, Farahani J, Qiang Z (2010) Managing water in rainfed agriculture—the need for a paradigm shift, Agric Water Manag 97:543–550

Rodrigues M, Santos L, Tiago P, Rodrigues AP, de Souza CR, Lopes CM, Maroco JP, Pereira JS, Chaves MM (2008) Hydraulic and chemical signalling in the regulation of stomatal conductance and plant water use in field grapevines growing under deficit irrigation. Funct Plant Biol 35:565–579

Romero P, Botía P (2006) Daily and seasonal patterns of leaf water relations and gas exchange of regulated deficit-irrigated almond trees under semiarid conditions. Environ Exp Bot 56:158–173

Romero P, Botía P, García F (2004) Effects of regulated deficit irrigation under subsurface drip irrigation conditions on vegetative development and yield of mature almond trees. Plant Soil 260:169–181

Romero R, Muriel JL, García-Tejero I (2009) A simple soil-plant-atmosphere model in simulink for irrigation control testing. Books of Abstracts ICHS. Viña del Mar, Chile, p 116, Nov 2009

Roose E (1996) Land husbandry—components and strategy. 70 Soils Bull. www.fao.org/docrep/T1765E/T1765E00.htm

Roose E, Kabore V, Guenat C (1993) Le zaï: Fonctionnement, limites et ame' lioration d'une pratique traditionnelle africaine de rehabilitation de la vegetation et de la productivité des terres degrades en region soudano-sahelienne (Burkina Faso). Cah Orstom, sér, Pédiol 2:159–173

Rosegrant M, Ximing C, Cline S, Nakagawa N (2002) The role of rainfed agriculture in the future of global food production. EPTD Discussion Paper No. 90, Environment and production technology Division, IFPRI, Washington (p 105). http://www.ifpri.org/divs/eptd/dp/papers/eptdp90.pdf

Rosengrant MK, Cai X, Cline SA (2002) World Water and Food to 2025. International Food Policy Research Institute, Washington

Rosenzweig C, Iglesias A, Yang X, Epstein P, Chivian E (2000) Climate change and US Agriculture: the impacts of warming and extreme weather events on productivity, plant diseases and pests. Boston, MA. Centre for Health and the Global Environment, Harvard Medical School. http://www.med.harvard.edu/chge/reports/climate_change_us_ag.pdf

Rosenzweig C, Strzepek KM, Major DC, Iglesias A, Yates DN, McCluskey A, Hillel D (2004) Water resources for agriculture in a changing climate: international case studies. Global Environ Change 14:345–360

Rousseaux CM, Figuerola IP, Correa TG, Searles SP (2009) Seasonal variations in sap flow and soil evaporation in an olive (Olea europaea L.) grove under two irrigation regimes in an arid region of Argentina. Agric Water Manag 96:1037–1044

Ruiz SMC, Torrecillas A, Pérez PA, Domingo R (2000) Regulated deficit irrigation in apricot trees. Acta Hortic 537:759–766

Sagnard FM, Guilbert V, Fauchard C (2009) In situ characterization of soil moisture content using a monopole probe. J Appl Geophys 68:182–193

Saiko TA, Zonn IS (2000) Irrigation expansion and dynamics of desertification in the Circum-Aral region of Central Asia. Appl Geogr 20:349–367

Salisbury FB, Ross CW (1985) Plant physiology. Wadsworth Publishing Company, Belmont, pp 428–429

San José J, Montes R, Nikonova N (2007) Seasonal patterns of carbon dioxide, water vapour and energy fluxes in pineapple. Agric Forest Meteo 147:16–34

Sánchez BMJ, Torrecillas A, León A, del Amor F (1989) The effect of different irrigation treatments on yield and quality of Verna lemon. Plant Soil 120:299–302

Sánchez PJM, Antiguedad I, Arrate I, García LC, Morell I (2003) The influence of nitrate leaching through unsaturated soil on groundwater pollution in an agricultural area of the Basque country: a case study. Sci Total Environ 317:173–187

Sánchez BP, Egea I, Romojaro F, Martínez MMC (2008) Sensorial and chemical quality of electron beam irradiated almonds (*Prunus amygdalus*). Food Sci Technol 41:42–449

Santos TP, Lopes CM, Rodrigues ML, Souza CR, Maroco JP, Pereira JS (2003) Partial rootzone drying: Effects on growth and fruit quality of field-grown grapevines (*Vitis vinifera* L.). Funct Plant Biol 30:663–671

Schaffer B, Whyley A (2002) Enviromental physiology. In: Whiley A, Shaffer B, Wolstenholme B (eds) The avocado Botany production and uses. CABI Publisching, Wallingford, pp 135–160

Schiettecatte W, Ouessar M, Gabriels D, Tanghe S, Heirman S, Abdelli F (2005) Impact of water harvesting techniques on soil and water conservation: a case study on a micro catchment in southeastern Tunisia. J Arid Environ 61:297–313

Schlesinger WH (1997) Biogeochemistry: an analysis of global change. Academic Press, San Diego

Scholander PF, Hammel HT, Hemingsen EA, Bradstreet ED (1964) Hydrostatic pressure and osmotic potential of leaves of mangrove and some other plants. Proc Natl Acad Sci USA 52: 119–125

Schröter D, Cramer W, Leemans R, Prentice C, Araújo MB, Arnell NW, Bondeau A, Bugmann H, Carter TR, Gracia CA, De la Vega LAC, Erhard M, Ewert F, Glendining M, House JI, Kankaanpää S, Klein RJT, Lavorel S, Lindner M, Metzger MJ, Meyer J, Mitchell TD, Reginster I, Rounsevell M, Sabaté S, Sitch S, Smith B, Smith J, Smith P, Sykes MT, Thonicke K, Thuiller W, Tuck G, Zaehle S, Zierl B (2005) Ecosystem Service Supply and Vulnerability to Global Change in Europe. Science 310:1333–1337

Scott CA, Shah T (2004) Groundwater overdraft reduction through agricultural energy policy: insights from India and Mexico. Water Res Develop 20:149–164

Seckler D (1996) The new era of water resources management: from dry to wet water savings. Res Rpt 1. International Irrigation Management Institute, Colombo, Sri Lanka

Seckler D, Molden DJ, Sakhtivadivel R (2003) The concept of efficiency in water resources management and policy. In: Kijne JW, Barker R, Molden DJ (eds) Water productivity in agriculture: limits and opportunities for improvement. CABI with IWMI, Wallingford

Segelken R (1997) US could feed 800 million people with grain that livestock eat. http://www.news.cornell.edu/releases/Aug97/livestock.hrs.html

Sentek (1999) Diviner 2000. Guia del usuario. Versión 1.0

Sentek (2001) Calibration of Sentek Pty Ltd Soil Moisture sensors. Sentek Pty Ltd, Stepney, Australia

Sevanto S, Höltta T, Nikinmaa E (2009) The effects of heat storage during low flow rates on the output of Granier-Type Sap-Flow Sensors. Acta Hortic 846:45–52

Seyfried MS, Hanson CL, Murdock MD, Van Vactor S (2001) Long-term lysimeter database, Reynolds creek experimental watershed, Idaho, United States. Water Resour Res 37:853–2856

Shackel KA, Ahmadi H, Biasi W, Bucner R, Goldhamer D, Gurusinghe S, Hasey J, Kester D, Krueger B, Lampinen B, McGourty G, Micke W, Mitcham E, Olson B, Pellettrau K, Philips H, Tamos D, Schwankl L, Sibbet S, Snyder R, Soutwick S, Stevenson M, Thorpe M, Weinbaum S, Yeager J (1997) Plant water status as an index of irrigation need in deciduous fruit trees. Hortic Tech 7:23–29

Shackel KA, Lampinen B, Sibbet S, Olson W (2000) The relation of midday stem water potential to the growth and physiology of fruit trees under water limited conditions. Acta Hortic 537:425–430

Shahin M (1996) Hydrology and scarcity of water resources in the Arab region. IHE Monograph 1. A.A. Balkema, Rotterdam, Brookfield, p 137

Shahnazari A, Liu F, Andersen MN, Jacobsen S, Jensen CR (2007) Effects of partial root-zone drying (PRD) on yield, tuber size and water use efficiency in potato (*Solanum tuberosum*) under field conditions. Fields Crops Res 100:117–124

Shepherd MA, Chambers B (2007) Managing nitrogen on the farm: the devil is in the detail. J Sci Food Agric 87:558–568

Shepherd MA, Hatch DJ, Jarvis SC, Bhogal A (2001) Nitrate leaching from reseeded pasture. Soil Use Manag 17:97–105

Shideed K, Oweis T, Gabr M, Osman M (2005) Assessing on-farm water-use efficiency: a new approach. ICARDA/ESCWA editions, p 86

Shiklomanov IA (1993) World fresh water resources. In: Gleick P (ed) Water in crisis: a guide to the world's fresh water resources. Oxford University Press, Oxford, pp 13–24

Shiklomanov IA (2000) Appraisal and assessment of world water resources. Water Int 25:11–32

Shiklomanov IA, Rodda JC (2003) World water resources at the beginning of the twenty-first century. Cambridge University Press, Cambridge

Sinclair TR, Tanner CB, Bennett JM (1984) Water-use efficiency in crop production. Bioscience 34:36–40

Singh AK, Kumar A, Singh KD (1998) Impact of rainwater harvesting and recycling on crop productivity in semi-arid areas—a review. Agric Rev 19:6–10

Singh Y, Singh RS, Lal RP (2010) Deficit irrigation and nitrogen effects on seed cotton yield, water productivity and yield response factor in shallow soils of semi-arid environment. Agric Water Manag 97:965–970

Smedema LK (1990) Irrigation performance and waterlogging and salinity. Irrig Drain Syst 4:367–374

Smil V (1999) Nitrogen in crop production: an account of global flows. Global Biogeochem Cycl 13:647–662

Smit B, Smithers J (1994) Sustainable agriculture: interpretations, analyses and prospects. Can J Reg Sci 16:499–524

Smit B, Harvey E, Smithers C (2000a) How is climate change relevant to farmers?. In: Scott D, Jones B, Audrey J, Gibson R, Key P, Mortsch L, Warriner K (eds) 2000. In: Climate change communication: proceedings of an international conference. Kitchener-Waterloo, Environment Canada: Hull, Quebec, Canada

Smith DM, Allen SJ (1997) Measurement of sap flow in plant stems. J Exp Bot 47:1833–1844

Smith WK, Hollinger DY (1991) Measuring stomatal behaviour. In: Lassoie JP, Hinckley TM (eds) Techniques and approaches in forest tree ecophysiology. CRC Press, Boca Raton, pp 141–174

Smith KA, Chalmers AG, Chambers BJ, Christie P (1998) Organic manure phosphorus accumulation, mobility and management. Soil Use Manag 14:154–159

Smith TE, Grattan SR, Grieve CM, Poss JA, Suarez DL (2010) Salinity's influence on boron toxicity in broccoli: I. Impacts on yield, biomass distribution, and water use. Agric Water Manag 97:777–782

Smithers J, Blay PA (2001) Technology innovation as a strategy for climate adaptation in agriculture. Appl Geography 21:175–197

Snell A (1997) The effect of windbreaks on crop growth in the Atherton Tablelands of North Queensland. Rural Industries Research and Development Corporation. RIROC. www.rirdc.gov.au/pub/shortreps/sr67.html

Snyder R (2000) Measuring crop water use in California Rice—2000, Department of Land, Air and Water Resources, University of California (Davis). http://www.syix.com/rrb/00rpt/wateruse.htm

Solley WB, Pierce RR, Perlman HA (1998) Estimated use of water in the United States in 1995: U.S. Geological Survey Circular 1200, 71 p

Sophocleous M (2000) From safe yield to sustainable development of water resources—the Kansas experience. J Hydrol 235:27–43

Southworth J, Randolph JC, Habeck M, Doering OC, Pfeifer RA, Rao DG, Johnston JJ (2000) Consequences of future climate change and changing climate variability on maize yields in the Midwestern United States. Agr Ecosyst Environ 82:139–158

Spreer W, Nagle M, Neidhart S, Carle R, Ongprasert S, Müller J (2007) Effect of regulated deficit irrigation and partial rootzone drying on the quality of mango fruits (Mangifera indica L., cv. 'Chok Anan'). Agric Water Manag 88:173–180

Spreer W, Ongprasert S, Hegele M, Wünnsche J, Müller J (2009) Yield and fruit development in mango (*Mangifera indica* L. cv. Chok Anan) under different irrigation regimes. Agric Water Manag 96:574–584

Stamatiadis S, Liopa-Tsakalidi A, Maniati LM, Karageorgou P, Natioti E (1996) A Comparative study of soil quality in two vineyards differing in soil management practices. In: Doran JW, Jones AJ (eds) Methods for assessing soil quality, Special publication N° 49, Soil Sci Soc Am, Madison, WI, USA, pp 381–392

Steduto P, Katerji N, Puertos MH, Unlu M, Mastrorilli M, Rana G (1997) Water-use efficiency of sweet sorghum under water stress conditions gas-exchange investigations at leaf and canopy scales. Field Crops Res 54:221–234

Stenitzer E (1993) Monitoring soil moisture regimes of field crops with gypsum blocks. Theor Appl Climatol 48:159–165

Stewart JI (1988) Response farming in rainfed agriculture. The WHARF Foundation Press, Davis, p 103

Stroosnijder L (2003) Technologies for improving rain water use efficiency in semiarid Africa. In: Proceedings of the symposium and workshop on water conservation technologies for sustainable dryland agriculture in Sub-Saharan Africa, Bloemfontein, South Africa, April 8–11, 2003. Pretoria, South Africa. ARC Institute for Soil, Climate and Water, pp 60–74

Suarez D (1992) Perspective on irrigation management and salinity. Outlook Agric 21:287–291

Suleiman AA, Tojo SCM, Hoogenboom G (2007) Evaluation of FAO-56 crop coefficient procedures for deficit irrigation management of cotton in a humid climate. Agric Water Manag 91:33–42

Szabolcs I (1996) An overview of soil salinity and alkalinity in Europe. In: Misopolinos N, Szabolcs I (eds) Soil salinization and alkalization in Europe. European Society for Soil Conservation, Giahudis Giapulis, pp 1–12

Tabor JA (1995) Improving crop yields in the Sahel by means of water-harvesting. J Arid Environ 30:83–106

Taiz L, Zeiger EE (1998) Plant physiology, 2nd edn. Sinauer Associates, Sunderland, p 792

Tayel MY, Ebtisam I, El-dardiry A, El-Hady M (2006) Water and fertilizer use efficiency as affected by irrigation methods. Am Euroasian J Agric Environ Sci 1:294–301

Tennakoon SB, Milroy SP (2003) Crop water use and water use efficiency on irrigated cotton farms in Australia. AgricWater Manag 61:179–194

Therios IN (2009) Olives. In: Atherton J, Rees A (eds) Irrigation of the olive crop production science in horticulture series. CAB International Wallingford, Osfordshire, pp 151–166

Thierfelder C, Wall PC (2009) Effects of conservation agriculture techniques on infiltration and soil water content in Zambia and Zimbabwe. Soil Till Res 105:217–227

Thirgood JV (1981) Man and the Mediterranean Forest a history of resource depletion. Academic Press, London, p 199

Thomas RJ (2008) Opportunities to reduce the vulnerability of dryland farmers in Central and West Asia and North Africa to climate change. Agric Ecosyst Environ 126:36–45

Thomas DSG, Middleton NJE (1994a) Desertification, exploding the myth. Wiley, Chichester

Thomas DSG, Middleton NJ (1994b) Salinization: new perspectives on a major desertification issue. J Arid Environ 24:95–105

Thrupp LA (2000) Linking agricultural biodiversity and food security: the valuable role of agrobiodiversity for sustainable agriculture. Int Aff 76:283–297

Tian Y, Su D, Li F, Li X (2003) Effect of rainwater harvesting with ridge and furrow on yield of potato in semiarid areas. Field Crops Res 84:385–391

Tilman D, Cassman KG, Matson PA, Naylor R, Polasky S (2002) Agricultural sustainability and intensive production practices. Nature 418:671–677

Tognetti R, d'Andrina R, Morelli G, Alvino A (2005) The effect of deficit irrigation on seasonal variations of plant water use in *Olea europea* L. Plant Soil 273:139–155

Torrecilla NJ, Galvez JP, Zaera LG, Retarnar JF, Alvarez A (2005) Nutrient sources and dynamics in a Mediterranean fluvial regime (Ebro river, NE Spain) and their implications for water management. J Hydrol 304:166–182

Torrecillas A, Ruiz-Sánchez MC, León A, Del Amor F (1989) The response of young almond trees to different drip-irrigated conditions—development and yield. J Hort Sci 64:1–7

Torrecillas A, Domingo R, Galego R, Ruiz-Sánchez MC (2000) Apricot tree response to withholding irrigation at different phenological periods. Sci Hort 85:201–215

Trisorio LG, Hamdy A (2008) Rain-fed agriculture improvement: water management is the key challenge. World Water Congress, Montpellier, p 12

Tuijl W (1993) Improving water use in agriculture: experience in the Middle East and North Africa. World Bank, Washington

Turner NC (1988) Measurements of plant water status by the pressure chamber technique. Irrig Sci 9:289–308

Turner NC (2004a) Sustainable production of crops and pastures under drought in a Mediterranean environment. Ann Appl Biol 144:139–147

Turner NC (2004b) Agronomic options for improving rainfall-use efficiency of crops in dryland farming systems. J Exp Bot 55:2413–2425

Turral H, Svendsen M, Faures JM (2010) Investing in irrigation: Reviewing the past and looking to the future. Agric Water Manag 97:551–560

UN (2004) World population prospects: the 2004 revision population database. United Nations Population Division, Washington DC, USA

UN (2007) Unite Nations, Press Release POP/952, Department of public information, News and media division, NY, USA. http://www.un.org/News/Press/docs//2007/pop952.doc.htm

UNCCD (2001) Measures to combat desertification and mitigate the effects of drought. Technologies for combating desertification India. National Programme to Combat Desertification. Ministry of Environment and Forests, Government of India, New Delhi, India

UNCSD (1997) United Nations Commission on Sustainable Development (UNCSD): Comprehensive assessment of the freshwater resources of the world, Report E/CN.17/1997/9. http://www.un.org/esa/sustdev/sdissues/water/waterdocuments.htm

UNEP (1991) Status of desertification and implementation of the United Nations Plan of action to the combat desertification. United Nations Environment Programme, Nairobi

UNEP (2003) Water Scarcity in the Middle East-North African Region. http://www.skmun. freeservers.com/unep/unepres1.htm

UNESCO (2001) Securing the food supply. World Water Assessment Programme, United Nations. United Nations Education Scientific and Cultural Organization, Paris, France

USBC (2001) Statistical Abstract of the United States 2001. U.S. Bureau of the Census, U.S. Government Printing Office, Washington DC

USDA (1997) Farm and Ranch Irrigation Survey (1998) 1997 Census of Agriculture, vol 3, Special Studies, Part 1, p 280

USDA (2001) Agricultural Statistics, Washington DC, USA

USDA-NASS (1998) Farm and Ranch Irrigation Survey, Census of Agriculture 1997, Estimated Quantity of Water Applied and Method of Distribution by Selected Crop: 1998 and 1994. U.S. Department of Agriculture, National Agricultural Statistics Service. http://www.nass.usda. gov/census/census97/fris/tbl23.pdf

USDA-NASS (2002) U.S. Department of Agriculture, National Agricultural Statistics Service. Washington DC, USA

USGS (2003) Water use. http://wwwga.usgs.gov/edu/wugw.html

Van Asten PJA, Barro SE, Wopereis MCS, Defoer T (2004) Using farmer knowledge to combat low productive spots in rice fields of a Sahelian irrigation scheme. Land Degr Develop 15:383–396

Van der Zee SEATM, Boesten JJTI (1993) Pesticide leaching in heterogeneous soils with oscillating flow: approximations. Sci Total Environ 132:167–179

Van Dijck SJE, Van Asch TWJ (2002) Compaction of loamy soils due to tractor traffic in vineyards and orchards and its effect on infiltration in southern France. Soil Till Res 63:141–153

Van Dijk J, Ahmed MH (1993) Opportunities for expanding water harvesting in sub-Saharan Africa: the case of the Teras of Kassala, Gatekeeper, vol Series 40. International Institute for Environment and Development, London

Van Duivenbooden N, Pala M, Studer C, Bielders CL, Beukes DJ (2000) Cropping systems and crop complementarity in dryland agriculture to increase soil water use efficiency: a review. Neth J Agric Sci 48:213–236

Vélez JE (2004) Programación del riego en cítricos en base a sensores de medida del estado hídrico del suelo y la planta. Ph.D. Thesis. Universidad Politécnica de Valencia, Spain, p 113

Verhoef A, Fernández GJ, Diaz EA, Main BE, El-Bishti M (2006) The diurnal course of soil moisture as measured by various dielectric sensors: Effects of soil temperature and the implications for evaporation estimates. J Hydrol 321:147–162

Verreynne JS, Rabe F, Theron KI (2001) The effect of combined deficit irrigation and summer trunk girdling on the internal fruit quality of 'Marisol' Clementines. Sci Hortic 91:25–37

Villalobos FJ, Fereres E (1990) Evaporation measurements beneath corn, cotton and sunflower canopies, Agron J 1153–1159

Vohland K, Barry B (2009) A review of in situ rainwater harvesting (RWH) practices modifying landscape functions in African drylands. Agric Ecosyst Environ 131:119–127

Vorosmarty CJ, Green P, Salisbury J, Lammers RB (2000) Global water resources: vulnerability from climate change and population growth. Science 289:284–288

Wahbi S, Wakrim R, Aganchich B, Tahi H, Serraj R (2005) Effects of partial rootzone drying (PRD) on adult olive tree (Olea europea) in field conditions under arid climate. I. Physiological and agronomic response. Agric Ecosyst Environ 106:289–301

Wakindiki IIC, Ben-Hur M (2004) Indigenous soil and water conservation techniques: effects on runoff, erosion, and crop yields under semi-arid conditions. Aust J Soil Res 40:367–379

Wall GJ, Pringle EA, Sheard RW (1991) Intercropping red clover with silage corn for soil erosion control. Can J Soil Sci 71:137–145

Wallace JS (2000) Increasing agricultural water use efficiency to meet future food production. Agric Ecosyst Environ 82:105–119

Wallace JS, Batchelor CH (1997) Managing water resources for crop production. Philos Trans R Soc Bio Sci 352:937–947

Wang CY, Li AIM, Wang DL (2007a) Growth and eco-physiological performance of cotton under water stress conditions. Agric Sci China 6:949–955

Wang D, Kang Y, Wan S (2007b) Effect of soil matric potential on tomato yield and water use under drip irrigation condition, Agric Water Manag 87:180–186

Wang FX, Kang Y, Liu SP, Hou XY (2007c) Effects of soil matric potential on potato growth under drip irrigation in the North China Plain. Agric Water Manag 88:34–42

Wani SP, Pathak P, Sreedevit TK, Singh HP, Singh P (2003) Efficient management of rain water for increased crop productivity and groundwater recharge in Asia. In: Kijne JW, Barket R, Molden D (eds) Water productivity in agriculture: limits and opportunities for improvement. CABI Publishing and International Water Management Institute, Wallingford

Wanjura DF, Hatfield JL, Upchurch DR (1990) Crop water stress index relationships with crop productivity. Irrig Sci 11:93–99

Warren A (1995) Comments on conservation, reclamation and grazing in the northern Negev: contradictory or complementary concepts? http://www.odi.org.uk/pdn/papers/38a.pdf

Water Framework Directive (2000) 2000/60/EC, OJ L327 1–72, 2000

WCD (2000) Dams and development. A new framework for decision making, The report of the World Commission on Dams, Earth Scan, UK

Webster GP, Poulton PR, Goulding KWT (1999) Nitrogen leaching from winter cereals grown as part of a 5-year ley–arable rotation. Eur J Agron 10:99–109

Wichelns D (2002) An economic perspective on the potential gains from improvements in irrigation water management. Agric Water Manag 52:233–248

Wichelns D, Oster JD (2006) Sustainable irrigation is necessary and achievable, but direct costs and environmental impacts can be substantial. Agric Water Manag 86:114–127

Wilson KB, Hanson PJ, Mulholland PJ, Baldocchi DD, Wullschleger SD (2001) A comparison of methods for determining forest evapotranspiration and its components: sap-flow, soil water budget, eddy covariance and catchment water balance. Agric Meteorol 106:153–168

Woldearegay K (2002) Surface water harvesting and groundwater recharge with implications to conjunctive water resource management in arid to semi-arid environments (with a model site of the Mekelle area, northern Ethiopia), Addis Ababa, Ethiopia

Wood S, Sebastian K, Scherr SJ (2000) Pilot analysis of global ecosystems–Agroecosystems. International Food Policy Research Institute (IFPRI) and World Resources Institute (WRI), Washington DC

World Bank (2003) World Development Report 2003: sustainable development in a dynamic world-transforming institutions, growth, and quality of life. The World Bank, Washington DC

Woyessa YE, Pretorius E, van Heerden PS, Hensley M, Van Rensburg LD (2005) Implications of rainwater harvesting in a river basin management: evidence from the Modder Basin, South Africa. Trans Ecol Environ 83:257–266

Wriedt G, Van der Velde M, Aloe A, Bouraoui F (2009) Estimating irrigation water requirements in Europe. J Hydrol 373:527–544

Yang Y, Watanabe M, Zhang X, Zhang J, Wang Q, Hayashi S (2006) Optimizing irrigation management for wheat to reduce groundwater depletion in the piedmont region of the Taihang Mountains in the North China Plain. Agric Water Manag 82:25–44

Youngquist W (1997) GeoDestinies: the inevitable control of earth resources over nations and individuals. National Book Company, Portland

Zaal F (2002) Explaining a miracle: intensification and the transition towards sustainable small-scale agriculture in dryland Machakos and Kutui districts Kenya. World Develop 30:1271–1287

Zacharias I, Koussouris T (2000) Sustainable water management in the European Islands. Phys Chem Earth 25:233–236

Zalidis G, Gerakis A, Misopolinos N, Prodromou K, Apostolakis A (1999) The impact of soil and water resources management on salt accumulation in Greece. In: Leone AP, Steduto P (eds) Salinity as a limiting factor for agricultural productivity in the Mediterranean Basin. Proceedings of the first trans-national meeting, Naples, Italy, pp 87–95

Zegbe DJA, Behboudian MH, Lang A, Clothier BE (2003) Deficit irrigation and partial rootzone drying maintain fruit dry mass and enhance fruit quality in 'Petopride' processing tomato (*Lycopersicon esculentum* Mill.). Sci Hortic 98:505–510

Zeitouni N, Dinar A (1997) Mitigating negative water quality and quality externalities by joint management of adjacent aquifers. Environ Resour Econ 9:1–20

Zhang J, Davies WJ (1989) Abscisic acid produced in dehydrating roots may enable the plant to measure the water status of the soil. Plant Cell Environ 12:73–81

Zhang J, Davies WJ (1990) Changes in the concentration of ABA in xylem sap as a function of changing soil water status will account for changes in leaf conductance. Plant Cell Environ 13:277–285

Zhang H, Oweis T (1999) Water–yield relations and optimal irrigation scheduling of wheat in the Mediterranean region. Agric Water Manag 38:195–211

Zhang J, Schurr U, Davies WJ (1987) Control of stomatal behaviour by abscisic acid which apparently originates in roots. J Exp Bot 38:1174–1181

Zhang Z, Wei X, Li X, Wang X, Xie Z (2004) Analysis on investment and benefit of harvested rainwater utilization in the northwest loess Plateau. Adv Water Sci 15:813–818

Zhang B, Min LF, Huang G, Cheng ZY, Zhang Y (2006) Yield performance of spring wheat improved by regulated deficit irrigation in an arid area. Agric Water Manag 79:28–42

Zhang X, Chen S, Sun H, Pei D, Wang Y (2008) Dry matter, harvest index, grain yield and water use efficency as affected by water supply in winter wheat. Irrig Sci 27:1–10

Zhi M (2000) Water efficient irrigation and environmentally sustainable irrigated rice production in China. International Commission on Irrigation and Drainage, Department of Irrigation, Wuhan University, China. http://www.ICID.org./wat_Mao.pdf

Zoebl D (2006) Is water productivity a useful concept in agricultural water management? Agric Water Manag 84:265–273

Zougmoré R, Zida Z, Kamboua NF (2003) Role of nutrient amendments in the success of half-moon soil and water conservation practice in semiarid Burkina Faso. Soil Till Res 71:143–149